T0275910

Grand Challenges
for Science
in the
21st Century

Exploring Complexity

For four centuries our sciences have progressed by looking at its objects of study in a reductionist manner. In contrast complexity science, that has been evolving during the last 30–40 years, seeks to look at its objects of study from the bottom up, seeing them as systems of interacting elements that form, change, and evolve over time. Complexity therefore is not so much a subject of research as a way of looking at systems. It is inherently interdisciplinary, meaning that it gets its problems from the real non-disciplinary world and its energy and ideas from all fields of science, at the same time affecting each of these fields.

The purpose of this series on complexity science is to provide insights in the development of the science and its applications, the contexts within which it evolved and evolves, the main players in the field and the influence it has on other sciences.

For the complete list of volumes in this series, please visit www.worldscientific.com/series/ec

Exploring Complexity — Volume 7

Grand Challenges for Science in the 21st Century

Editors

Jan W. Vasbinder
Para Limes at Nanyang Technological University, Singapore

Balázs Gulyás
Jonathan Y. H. Sim
Nanyang Technological University, Singapore

NEW JERSEY · LONDON · SINGAPORE · BEIJING · SHANGHAI · HONG KONG · TAIPEI · CHENNAI · TOKYO

Published by

World Scientific Publishing Co. Pte. Ltd.
5 Toh Tuck Link, Singapore 596224
USA office: 27 Warren Street, Suite 401-402, Hackensack, NJ 07601
UK office: 57 Shelton Street, Covent Garden, London WC2H 9HE

Library of Congress Cataloging-in-Publication Data
Names: Gulyás, Balázs, editor. | Vasbinder, Jan W., editor. | Sim, Jonathan, editor. |
    Nanyang Technological University, host institution.
Title: Grand challenges for science in the 21st century / edited by Balázs Gulyás,
    Jan Vasbinder, Jonathan Sim.
Description: New Jersey : World Scientific, [2018] | Series: Exploring complexity ; volume 7 |
    Includes bibliographical references and index.
Identifiers: LCCN 2018036134 | ISBN 9789813276437 (hardcover : alk. paper) |
    ISBN 9789813276680 (softcover : alk. paper)
Subjects: LCSH: Science--Social aspects.
Classification: LCC Q175.5 .G732 2018 | DDC 303.48/3--dc23
LC record available at https://lccn.loc.gov/2018036134

British Library Cataloguing-in-Publication Data
A catalogue record for this book is available from the British Library.

For any available supplementary material, please visit
https://www.worldscientific.com/worldscibooks/10.1142/11161#t=suppl

Printed in Singapore

Contents

Acknowledgements

This book would not have been possible if not for Miss Ngiam Liyi and Miss Yuka Kamamoto, who tirelessly transcribed several hours of recorded presentations and discussions by our speakers, thereby laying the groundwork for the entire editing process.

We would like to extend our gratitude and thanks to Mr. Isaac Tan and Mr. Samuel Chong, who spent many hours proofreading the manuscripts, and helping us to correct typos and errors every step of the way. Special mention should be given to Mr. Samuel Chong, for his incredible research skills in locating the many academic references to books and journal papers that were raised by our speakers during the presentations and discussions. We have included these citations in the footnotes for your reference and convenience.

List of Speakers

W. Brian Arthur

External Professor, Santa Fe Institute

Visiting Researcher, System Sciences Lab, Palo Alto Research Centre (PARC)

Brian Arthur is an economist and one of the pioneers in the science of complexity. He received his PhD in Operations Research from Berkeley, and has many other degrees in economics, electrical engineering, and mathematics. For many years, he was at Stanford University before moving on to the Santa Fe Institute (SFI), where he continues to be a member of the SFI Founders Society. He has served on the SFI's Science Board for 18 years, and its Board of Trustees for 10 years.

In 1988, Arthur directed SFI's first research programme, which later became the basis for Complexity Economics, which he proposed as an alternative to the classical approach to economics. He pioneered the modern study of "positive feedbacks", or "increasing returns" in the economy — their role in amplifying small random events in the economy and locking in dominant players. This seminal work has become the basis for our contemporary understanding of the high-tech economy.

Sydney Brenner

Senior Fellow, Agency for Science, Technology, and Research (A*STAR), Singapore

Senior Fellow, Janelia Farm Research Campus, Howard Hughes Medical Institute (HHMI)

Adjunct Professor, Lee Kong Chian School of Medicine, Nanyang Technological University

Sydney Brenner received his education in South Africa, graduating with a degree in Biological Sciences and Medicine. In 1954, he received his doctorate in Physical Chemistry from Oxford University. His early research was in molecular genetics, where he is well known for his research in the 1960s, when he discovered messenger RNA (mRNA), together with Francois Jacob and Matthew Meselson.

In recognition of his research, Brenner has received many awards. He is a Fellow of the Royal Society (1965), a Foreign Associate of the US National Academy (1977), External Scientific Member of the Max-Planck Society (1988), and Associe Etranger, Academie des Sciences, Paris (1992). He received the Lasker Prize twice (1971 and 2000); two Gairdner International awards (1978 and 1991); and the Nobel Prize in Physiology of Medicine (2002) with H. Robert Horvitz and John E. Sulston. He is a Companion of Honour of the United Kingdom. And in 2017, Brenner was conferred the Grand Cordon of the Order of the Rising Sun by the Government of Japan, recognizing his instrumental role in establishing the Okinawa Institute of Science and Technology Graduate University as an internationally reputed research facility.

Helga Nowotny

*Professor Emerita of Social Studies of Science,
ETH (Swiss Federal Institute of Technology) Zurich,
Switzerland*

*Visiting Professor, Nanyang Technological
University, Singapore*

Helga Nowotny holds a PhD in Sociology from Columbia University, and a doctorate in jurisprudence from the University of Vienna, and she has since taught and held research positions in many esteemed institutions around the world. She has published widely in the field of Science and Technology Studies, and on social time.

Throughout her professional career Nowotny has been engaged in science and innovation policy matters and is serving as an advisor at the national and EU level. She is a founding member of the European Research Council (ERC). She was elected Vice President of the ERC in 2007, and later President of the ERC from 2010 to 2013. She is currently the chair of the European Research Area (ERA) Council Forum Austria, a member of the Austrian Council and Vice-President of the Council for the Lindau Nobel Laureate Meetings. She is also a Foreign Member of the Royal Swedish Academy of Sciences and continues to serve on many international advisory boards in Austria and throughout Europe.

Martin J. Rees

Fellow of Trinity College, and Professor Emeritus of Cosmology and Astrophysics, University of Cambridge

Martin Rees is a leading astrophysicist and a senior figure of the sciences in the United Kingdom. He holds the honorary title of Astronomer Royal, and was the President of the Royal Society (2005-2010). He has published widely, authoring over 500 scientific papers on astrophysics and cosmology, and several books. He has also contributed numerous magazine and newspaper articles on scientific and general subjects, and has broadcasted and lectured widely. For his services to the sciences, Rees was knighted in 1992, and later elevated to the title of Baron Rees of Ludlow in 2005.

Rees has done influential theoretical work on a diverse range of subjects, ranging from black hole formation to extragalactic radio sources. He was one of the first to predict the uneven distribution of matter in the Universe and proposed observational tests to determine the clustering of stars and galaxies. His greatest contribution to astrophysics is research focused on the so-called cosmic "dark ages" — the period shortly after the Big Bang when the universe was yet to have any source of light.

Terrence J. Sejnowski

Professor and Laboratory Head, Computational Neurobiology Laboratory, Howard Hughes Medical Institute Investigator, and Francis Crick Chair, The Salk Institute for Biological Studies

Terrence Sejnowski received his PhD in Physics from Princeton University, after which he was a postdoctoral fellow in the Department of Neurobiology at Harvard Medical School and on the faculty at Johns Hopkins University. Sejnowski is a pioneer in computational neuroscience and his goal is to understand the principles that link brain to behaviour. His laboratory uses both experimental and modelling techniques to study the biophysical properties of synapses and neurons and the population dynamics of large networks of neurons. He was instrumental in shaping the Brain Research through Advancing Innovative Neurotechnologies (BRAIN) Initiative that was announced by the White House in 2013, and served on the Advisory Committee to the Director, National Institutes of Health, for the BRAIN Initiative.

Sejnowski is the President of the Neural Information Processing Systems (NIPS) Foundation, which organizes an annual conference attended by over 2000 researchers in machine learning and neural computation. He was elected to the American Academy of Arts and Sciences. He is also a Fellow of the American Association for the Advancement of Science; a Fellow of the Institute of Electrical and Electronics Engineers; and a Fellow of the Cognitive Science Society. He has published over 500 scientific papers and 12 books, and has received many honours for his work, including the NSF Young Investigators Award, the Wright Prize for interdisciplinary research from Harvey Mudd College, the Neural Network Pioneer Award from the Institute of Electrical and Electronics Engineers, the Hebb Prize from the International Neural Network Society and the Rosenblatt Award from the Institute of Electrical and Electronics Engineers.

Eörs Szathmáry

Director, Parmenides Centre for the Conceptual Foundations of Science, Munich

Professor of Biology, Department of Plant Taxonomy and Ecology, Eötvös Loránd University, Budapest

Eörs Szathmáry is a Hungarian biologist whose primary interest resides in theoretical evolutionary biology. His research is focused on the common principles of the major steps in evolution, such as the origin of life, the emergence of cells, the origin of animal societies, and the appearance of human language. Together with his mentor, John Maynard Smith, he has published two important books which serve as the main references in the field: *The Major Transitions in Evolution* (Freeman, 1995), and *The Origins of Life* (Oxford University Press, 1999). Both books have been translated into several languages.

Szathmáry is on the editorial board of several journals, such as the *Journal of Theoretical Biology, Journal of Evolutionary Biology, Origins of Life and Evolution of the Biosphere, Evolutionary Ecology* and *Evolution of Communication*. He was awarded the New Europe Prize in 1996 by a group of institutes for advanced study. He used the prize to establish the New Europe School for Theoretical Biology (NEST) Foundation, whose task is to help young Hungarian theoretical biologists. He is also a member of Academia Europaea and the Hungarian Academy of Sciences.

Moderator

Tor Nørretranders

Acclaimed Writer and Thinker

Tor Nørretranders is an independent author, thinker and lecturer based in Copenhagen, Denmark, serving an international audience. Among the books he has authored, the best known are *The User Illusion* (Penguin, 1999), about consciousness and information; *The Generous Man* (Thunder's Mouth, 2005), about why humans dare, care and share in life; and *Commonities = Commons + Communities* (Samsoe Energiakademi, 2013), with Søren Hermansen.

Nørretranders writes on a vast range of issues, spanning from the philosophy of quantum mechanics, to the social networks of Civilisation 2.0. Sustainability has been a theme in his writings for many decades. His most recent book in Danish, *Vær nær*, deals with the emergence of societal cohesion through the unconscious bonding between individuals.

1. Introduction

In June 2016, Para Limes at Nanyang Technological University (NTU), Singapore, organised a four-day meeting — entitled *"Grand Challenges for Science in the 21st Century"* — with six renowned scientists to discuss the fundamental challenges that might shape the nature of science in the coming decades or even change the whole notion of science.

The scientist included the distinguished Nobel Laureate and pioneer in genetics and molecular biology Sydney Brenner, the leading economist and complexity thinker W. Brian Arthur, the highly influential astrophysicist, Astronomer Royal Martin J. Rees, the pioneer of neural networks and computational neurosciences Terrence J. Sejnowski, the world renowned evolutionary biologist Eörs Szathmáry, and the former president of the European Research Council and professor emeritus of social studies of science Helga Nowotny. The lively discussions were moderated by the Danish writer, Tor Nørretranders.

The videos of the presentations and discussions can be viewed on youtube under the link
https://www.youtube.com/watch?v=RjoN87LJmb0&index-=1&list=PLasWJveXPWTFCVG4GgzGiwvahR06bb8-I (Grand Challenges for Science in the 21st Century).

This book is a compilation of the edited transcripts of these videos. It is divided into two parts. The first part comprises the individual presentations of the scientists. The second part presents thematic roundtable discussions on issues that are of urgent concern to the world today.

Science is embedded in society which supports the activity of scientific exploration through public funding and which is altered and shaped by the insights, applications and technological developments arising from these explorations. We now live in a time when the rate of technological advancements far outpaces the capability of society to adapt to these

1

changes. This mismatch poses challenges for both science and society and disciplinary science cannot adequately address these challenges. In fact, many grand challenges arise from developments that were made possible by advances in disciplinary science, but that have never included the societal consequences of the application of these advances. By now we may have reached the point that for science to continue getting public support it must abandon its heavy disciplinary focus and accept its role as one of many agents in a very complex system.

In the light of these challenges it was fully appropriate that the six discussant scientists represented a broad spectre of sciences. Their discussions never got stuck in a disciplinary trap and covered the entire spectrum of science.

Jan Wouter Vasbinder
Balázs Gulyás
Jonathan Y. H. Sim

Part I

Part I

2. The Management of Science and the Need for Greater Clarity

Sydney Brenner

If you were to ask this general question to any scientist, certainly most of them in Singapore, America, and Europe, "What are the grand challenges of science?" I'll tell you what they will say.

They will say: "The grand challenge is: How can I get hold of money to do science?"

I think that in itself is an interesting development. Science is something now which countries practise and are proud of. They have to spend a certain fraction of their GNP on what they call, "Research and Development", and this is the way they are measured in the league throughout the world.

The working scientist's big challenge is: what can one do to get money to support one's work? If you observe how science has evolved since the beginning of the 20th century, you would find this to be a very interesting subject to study. That's because nowadays, most of the people who say they are in science aren't really in science. They are in something else. They are in the *management* of science! They are in something I call, "3M science" (3M is a great company, by the way). The 3Ms are: Money, Machines, and Management. These people believe that everything can be solved by the application of what the Americans call "process".

You simply need to have a way of doing something, apply the resources, and you will get the answer. They are not so worried about challenges such as the ones that some of us think about. Their only challenges are: Will I be awarded good points? Will I get promoted? Will I be able to survive in the economy of science?

In a funny way, one might ask, "Does science create the scientists, or do the scientists create science?"

Now, I think there are great challenges, and I think there have been, and still will be in the future. But of course, for a practicing scientist, it is important to know whether you can do anything about them, because action is one of the things that had distinguished, let's say, physics, from its old name of natural philosophy. Of course, biology was never called "natural engineering", but that is what it would be called if it had that.

Now, of course, one of the advantages of a scientist is that he is supposed to tell you the truth, the *real* truth. Of course, there are various shades of truth, but I think the important thing is, if we do things which have no experimental confirmation, how can we call that science? That's the Popperian argument. Karl Popper never believed that evolution was science.[a] And he probably didn't believe that cosmology was a science either. You can't disprove them with experiments. But of course, we don't believe that to be the case.

So there is real science to be practised, and what I think is very interesting, is that it can't be practised without having an army to support it, a whole expedition.

But what it also requires is clarity of thought, the clarity to discover ways of finding what's wrong with our present scientific endeavours. I think one of the really interesting things about trying to do science, is to carefully go over and over again what everybody says, and ask: "How much of this do we believe?"

I think that a shattering thing in the history of science is the amazing amount of nonsense that scientists will tolerate. They will tolerate it and are clearly hoping something will be done about it. It is tolerated, it is held, and the amazing thing is that it goes on like that, until one morning, you wake up and you say, "I don't like that story!", and you find there is something wrong with it. There is a little "lie" contained within the story, and it is from there that we can push that field of science further ahead.

---

[a] Ref. 1, pp. 343–345

Recently, I suddenly saw what was wrong with everything written on evolution since 1930. Everything is wrong, and we have to change it. There is an assumption about everything in evolution having only two fates: you either change, or you become extinct.

But why shouldn't there be a fate of just go on doing what you are doing if it's okay? There is no need to change if you are happy with that. So in a curious way, what I've discovered is that there are, not two, but three fates. And the third fate is: you don't change. You don't have to become extinct, you can just go on doing it. I think that everything in the world today is a living fossil. Crocodiles, even if they didn't like being crocodiles, seem to be quite comfortable lying there. Whatever they've done, they've done this for hundreds, thousands, millions of years! They still are crocodiles. They are living fossils of what's it like to be the same for thousands, millions of years.

The above example should give you a different view of things. We have tolerated the story of evolution for so long, without recognising the little "lie" in it. But as I have shown, when you have the clarity to find out what's wrong with it, you can push the story further. I think that is something that we can exploit. Of course, today, truth is not necessarily the way you get things published (in fact, sometimes, telling lies is a better way to get things published).

What we need is a kind of shift, where we're constantly questioning ourselves whether something is wrong. I am quite amazed with myself how much nonsense I've kept in my head without having the clarity to ask a different kind of question. I urge you to question things, to question the scientific "stories" around us. Because breaking the bonds that hold you are the important things in the grand challenges of science.

Reference

1. Popper, Karl. "Natural Selection and the Emergence of Mind." *Dialectica* 32, no. 3–4 (1978): 339–355.

3. The Potential of Algorithms

W. Brian Arthur

I remember reading a book in 1996 called, *The End of Science*. The writer, John Horgan, was arguing that the really big problems in science, or many of them, had been solved. The remaining big problems — Why do we exist at all? Why does anything exist? Where does consciousness come from? — would likely never be solved. So, science was at an end.[a]

I found this very puzzling. At that time, I was thinking of writing about technology, so Horgan's book caused me to think a lot.

What I want to talk about with regard to where science is heading, is to look at it from the point of view of technology, not just from the point of view of thought. We are trying to look 20, 30, 50 years ahead, and science proceeds in many ways theoretically, by thinking and by observing. But more than anything else, I would maintain, science proceeds by its technologies, its instruments, and the methods that are tied to those instruments.

I will come up fairly randomly with five or six different families of technology. One is very obvious. The optical telescope around 1610. Thereafter, the microscope 50 years later. And thereafter, particle accelerators. These are sort of random families of technology all through the 20th century and into our time.

X-ray crystallography was going in the 1930s and 40s. Radio astronomy was going from the 30s and 40s. Genomic methods started in 1953, but it got going in the 60s, 70s, and it is enjoying a real heyday at the moment.

---

[a] Ref. 1, pp. 6–7

(You might have your own list of technologies, and I have probably omitted at least a dozen really important families of technology.)

More importantly, what is common between them?

These technologies are instruments, and the methods that go with them keep improving. There is nothing static, and they improve over decades, and sometimes over centuries.

What I want to point out is, each of these technologies has opened up a new world for us. The telescope showed Galileo something really important. Venus had phases and was circling the Sun, which was an immense idea at the time. It showed Galileo that the moon was very imperfect. It was not a perfect body by any means. Above all, it showed him that Jupiter had moons circling it. The Earth was not the center, which Galileo probably knew from studying Copernicus. But this opened a whole new world that the Earth was not at the center of things and that there were amazing things going on in the cosmos.

I could take any of these methods, X-ray and crystallography, for example. Not only did it allow scientists to look into crystals, but it also allowed scientists to look into complicated polymers and DNA itself. That opened the world to molecular biology; to how genetics was fundamentally organized; to the work of Brenner, Crick, and others on the genetic code. It opened up the entire world of molecular biology. Radio astronomy, gave us quasars, pulsars, radio galaxies and many, many other things.

So it is not just that scientists or engineers come up with these instruments. If you noticed, these instruments derive from phenomena in nature, such as radio phenomena and optical phenomena. They reveal worlds to us that constitute the challenges of the times. It is a real challenge for Galileo, when he looked through his telescope, to see so much going on out there. And it took him and his successors decades and even centuries to work it out.

Is there a new technology or sets of new technologies at the moment besides the ones I have mentioned? Naturally, the one to point to is computation, and a naïve view of computation.

I work in Silicon Valley. I am at Xerox PARC. I have been there for a long time, at the Santa Fe Institute. We do an awful lot of computation. You might think that we are getting somewhere, probably because Moore's Law operates. There is so much more computational power available. There is more storage power available. There is more data available. So you might think that things are naturally moving ahead.

It is subtler than that. It is not really computation that counts (of course, you need the hardware). It is actually the algorithms that are making the difference, yet it is not just algorithms that count. But rather, it is the subtle interplay between technology or instruments themselves *and* the algorithms. For example, you could take certain optical instruments or sensors, and point them at something. You sense with these instruments. Yet, there is an awful lot of real-time computation, and they can compensate instantaneously for refractive changes in the atmosphere.

What you see again and again is a really interesting combination of instruments themselves looking at something in nature, digital versions of those instruments being brought back, algorithms somehow adjusting or magnifying those instruments in real time, and then we get some sort of feedback, and the whole thing is orchestrated as a whole and is far more than the original parts.

One way to think about this is that if you have algorithms, these algorithms can do much more than calculate. They can adjust, adapt, count, store, control, direct, recognize, compare, amplify, etc. Notice that all the descriptors I used are verbs: 'store', 'recognized', etc.

We have essentially invented for ourselves a new programming language in which we can take hardware instruments, put them together with these verbs ('amplify this', 'store that', 'retrieve this', 'sort this'), and we get extraordinarily powerful results. Not quite computers, not quite something like the Internet of Things, but new instruments that are adjusting themselves in real time.

Does it make a difference?

I want to quote some figures from Douglas Robertson who said optical telescopes made a difference to what we could see with our naked eye, originally by about a factor of sixty that goes up to about two orders of magnitude, some between a hundred and a thousand.[b] You might think that digital methods add 10% or 20% to the resolution of optical telescopes. No, it does not. It adds a further three to four orders of magnitude. So the computer or algorithms have added more to what we can see in the heavens than original telescopes have optically.[c]

And you could go across one field after another and say largely the same thing. What can we see with X-rays? Quite a bit. What can we see with computerized X-rays? Things like computerized tomography or CT scanning? Quite a lot more, and we can see in three dimensions. What can we see with magnetic resonance, with nuclear magnetic resonance? Not very much. But what can we see with computation allied to nuclear magnetic resonance? A huge amount! We can see soft tissues!

I could go through field after field after field, and the story is the same. We are seeing new worlds. We have enhanced understanding. With the computer, we can sit in a little capsule if we want and probe these new spaces, which might be quite theoretical.

Mathematicians of old were able to do a lot of calculations, and someone like Poincaré, a bit over a hundred years ago could do a lot of calculations by hand and figure out what might be happening. But it has been routine for the past 20 to 30 years, to use computers and see far more than Poincaré himself could have seen. Computation and algorithms are to the human mind what telescopes are to the eye. They are expanding the range of the human mind and now we have artificial intelligence, which is expanding things yet further. So the challenge is what are we going to see with all of that.

One of the things we are seeing is that a lot of the sciences I have been involved in, in Santa Fe and before that, have been looking at how systems unfold over time. The kinds of systems I was looking at were always linear. Generally speaking, they are in some form of equilibrium. They were somehow continuous and operated on continuous manifolds.

[b] Ref. 2, p. 20
[c] Ref. 2, p. 157

You could picture them as little ball bearings running down some kind of manifold, and maybe reacting to outside stimuli. What computation has allowed us to do is to reverse those four things. So we are starting to look at nonlinear systems, non-continuous ones, in other words, discrete systems, known equilibrium systems and systems that react to their own internal state, and how those changes in internal state play out over time in some larger sphere.

I was brought up to believe that linear equilibrium systems, continuous ones were the entire world, and then we are tiny little anomalies outside that. But what we are starting to realize (thanks to being able to compute our way to look at many more systems), is that most systems in nature are nonlinear. They are non-continuous. They are perpetually not in equilibrium and they are reacting to their own patterns. And this is opening up new worlds. I think this is a profound challenge.

Stanislaw Ulam in Los Alamos once said, "Long linear systems are to linear ones as non-elephants are to elephants."[d] There is a lot to the world of non-elephants, and we simply do not know it in what we are doing in science. We are starting to explore a way into these completely new worlds. Those new worlds do not look mechanistic, amazingly. They look organic, and they look messy, but a sort of messiness we are at home with. They are not super well-ordered, and they look alive. I think this is the main challenge that we are facing exploring all this.

References

1. Horgan, John. *The End of Science: Facing the Limits of Knowledge in the Twilight of the Scientific Age*. Reading, Mass: Addison-Wesley Pub, 1996.
2. Robertson, Douglas S. *Phase Change: The Computer Revolution in Science and Mathematics*. Oxford: Oxford University Press, 2003.
3. Campbell, David K. "Nonlinear Science from Paradigms to Practicalities." In *From Cardinals to Chaos: Reflections on the Life and Legacy of Stanislaw Ulam*, edited by Necia Grant Cooper. New York: Cambridge University Press, 1989.

---

[d] Ref. 3, p. 218

4. The Progress of Science and Artificial Intelligence from Efforts in Understanding the Brain

Terrence J. Sejnowski

Let me just start with a little story about the great catcher from the New York Yankees, Yogi Berra. He once said that it is very difficult to make predictions, especially about the future,[1] and that is the theme I wish to discuss.

I am going to pick a single problem to discuss. There was a Go game between a computer, *AlphaGo*, and Lee Sedol, a South Korean Go champion. Go is a very ancient and difficult game. It is played on a board with little pieces, and the idea is to get as much territory as possible. It is very subtle because it involves both tactics and strategy. Tactically, you need to figure out where to put your own pieces to surround the enemy, but strategically, you need to figure out what is going on around the whole board because it is very important how it plays out at the end of the game.

Lee Sedol was overconfident. He went into it knowing that there had been no Go computer game that had ever, ever come close to beating the equivalent of a grandmaster, a nine *dan* player, and certainly not a Go champion. He lost the first three games in a row and lost the match. What I want to do is open the hood and tell you how that happened.

AlphaGo was a program developed by DeepMind, a company in London, headed by Demis Hassabis, a card-carrying neuroscientist who had this idea that by using what we understand about how the brain functions, we can build a program that behaves like a brain but can overcome some of its limitations.

The key to how this program was able to reach human and superhuman levels of performance goes back to a very ancient part of the brain: the dopamine system. The idea here is that we have a very simple reinforcer that tells us about things and predicts future rewards. These dopamine neurons in the brainstem project very broadly throughout the entire cerebral cortex and the basal ganglia just beneath it. These are two very important parts of the brain that store knowledge. Everything you know about the world, your sensory perceptions, planning in the prefrontal cortex: that's all done by a hundred billion neurons throughout the cortical mantle. They all project down to the basal ganglia just beneath it. This is incredibly important for making decisions, taking actions, and computing the value of a particular action.

The dopamine neurons are signaling something that is absolutely essential to improving performance, and it is called: Reward Prediction Error. It is very simple, but it turns out to be a very powerful algorithm. If you get a reward, and more reward than you expect for an action, then the dopamine neurons elevate their firing rate, saying, "Wow, whatever you did was really good, do it again!" But if you get less reward for what you are doing, or if you are disappointed, the dopamine neurons lower their firing rate. They stop firing for a while, and they say, "Bad move, don't do that again." Really, what they are computing is potential for all future rewards into the distant future, discounted.

Reinforcement learning theory has been studied by psychologists and biologists for many decades. They call it classical conditioning: when you pair a sensory stimulus like a bell with a food reward, and then you begin to associate the bell with the reward.

It is a very simple paradigm. But we will see that it allows you to solve a very difficult problem, which is, "How do I go and make a sequence of decisions to get the reward?" Not just the last decision of hearing the bell, but how do I get to the point where I can hear the bell? I may have to make many decisions to get there, and in fact, you have had to make many decisions to get here in your life. Your decision to go to a certain school, your decision to learn a certain subject — all those things were preparatory for where you are now.

This is a very simple version of what we think is actually going on in your brain, and it is what was used by *AlphaGo*. Imagine there is a little corridor with a reward at the end of the turn. In order to get to your reward, you have got to figure out that you have to go to the end, turn right, and go down. Now, at the beginning, you are just randomly moving back and forth, but eventually you build up what is called the value function. The value function tells you what is the value of the move (for instance, at a particular point, it is much more valuable to move to the right than to the left). You sample all the different possibilities and you build up this value function. You will become very good at doing it automatically. You don't have to think about it anymore. This value function is what we build up through experience, over years and years of rewards and failures.

How do you compute the value function? Back in the 80s, we developed these very simple neural networks that took inputs, and through learning, produced an output. Back then, we could only deal with networks that had one layer of hidden units. That has all changed because computing has gotten so much faster, by a factor of a million over the last 30 years.

To show how this reinforcement learning algorithm can actually solve a difficult problem, here is an example. Gerald Tesauro, was at IBM's TJ Watson Research Center. He applied a similar algorithm called, Temporal Difference Learning, to the game of backgammon. Backgammon is a relatively simple game. There is an element of chance in it. You have to throw the dice to decide to go forward. I am not going to explain the game, except to say that there are strategies. It is basically a race to the finish, and you have to go through the enemy. Very subtle decisions have to be made about setting up a blocking position, and when you should run for it.

In any case, Gerald applied Temporal Difference Learning and had the program play against itself. That is a very crucial thing. If it played against a human, it would never be as good as that human, because the human is the teacher. By playing against itself, in principle, it can get better and better and better. There is no teacher, and so it is not limited by a teacher. It teaches itself.

After playing a few hundred thousand games, the program got better than Gerald. He had no idea how good it was, so he asked Bill Robertie, who has written books about backgammon, to come and play it. The program played 50 games, and was losing about a quarter a point per game, which against a really good, strong backgammon player, is pretty good. In fact, Robertie said, "This is really the best computer program I've ever played." He took notes of some of the moves, and went back and showed that even though no human would have made that move, it was actually the better move. It had discovered and created a new strategy. Others who had been hearing about this came. By this point, the program had gotten up to 800,000 games, and it was getting better and better and better. Finally, after more than a million and a half games, Robertie came back and it played him to a standstill. He said that at that point, this program could well beat the world champion.

This is a new form of artificial intelligence, because the program was learning through experience rather than being programmed by an engineer based on simple processing and rules, which is the tradition in AI. Somebody had to create the rules. Here, the program creates its own rules.

Where are we today?

AlphaGo had something else going for it: deep learning. Deep learning is based on what we know about the visual system, the cortical layers in the hierarchy. It starts with Hubel and Wiesel's classic work in 1962, where they show that unlike earlier stages in the retina, in the visual cortex they respond to linear features, like edges and bars.[a] Moreover, starting from the retina, there are 12 layers in the hierarchy, 24 different distinct visual topographic maps of the visual world in the cortex, which are highly interconnected. That turns out to also be an important ingredient. Once you reach into the higher levels of this hierarchy, the features that the cells respond to get more and more abstract. For example, instead of little edges, they respond to faces, or they will respond to a very complex scene where there is not just a single object, but several objects.

---

[a] Ref. 2, p. 151

This architecture of the visual system is reproduced in the deep learning architecture used to beat Lee Sedol in Go. There was a deep learning network that looked at the board and kept track of whether particular configurations of the board were good or bad. It learnt by playing against itself. That is the reinforcement learning algorithm.

And here is the beauty: because computers are so fast, it could play hundreds of millions of games. In fact, when they challenged Lee Sedol, they did not know how good *AlphaGo* was because they did not have any human that could play it at that level. *AlphaGo*'s win was really a shock to many, especially in Asia, where Go is a really big game. Lee Sedol is like a rock star in Korea. This was like a blow to humanity that *AlphaGo*'s win could be achieved. Go is a very complex game. Fortunately for us, Lee Sedol won the fourth game, although he lost the fifth. At least we are still in the game here.

I am going to end with saying that this story about *AlphaGo* and Lee Sedol is being recapitulated in many spheres at this very moment. I want to give a little brief introduction to a few of the examples. In 2005, Sebastian Thrun's car, which was equipped with sensors, won the DARPA Grand Challenge for autonomous vehicles.[3] His car was able to drive 200 miles over the desert and got to the finish line before many other competitors that were based on engineering practice, especially big trucks filled with super computers. Sebastian's vehicle was learning with the very same algorithms that was used by *AlphaGo*. The difference was that instead of programming rules, Sebastian drove the vehicle for thousands of miles through the desert. He did not know until the beginning of the race what the actual route would be. They were downloading the route literally moments before they had to take off. Before the race, Sebastian just rode through all the different types of routes in the desert, and the car would keep track of how he reacted to different sensory environments. If there was a rock, then Sebastian would avoid it. If there was a tumbleweed, he would go through it. The vehicle figured out from the different signatures, what to do under these different circumstances. Sebastian has been the head of Google's driverless car division, which has had more than a million miles of experience in the San Francisco Bay area driving a car without human intervention.

It is not just Google now. Apple is in the game. Uber announced that they are going into it. Even the auto companies have realized that they are going to fall behind if they don't start really picking up their pace, because this is probably where the future is going. This is a very disruptive technology, we cannot even imagine the impact it is going to have. Who could have imagined the impact that Amazon was going to have on marketing, Spotify on music, and Netflix on television?

There are unbelievable changes that are occurring and continuing to occur.

Here is the problem, going back to Yogi Berra. We cannot imagine the future. We are not very good at it. Introduce something new that is disruptive, and it is going to take a new dimension that we just won't be able to predict, or when we get there, be able to appreciate it.

This technology is accelerating. It is really quite remarkable. Predicting the future is very difficult.

There is a lot more complexity in the brain at many different levels of organization that are yet to be exploited. In 2013, the President at the White House made an announcement that the US was going to have a grand challenge for understanding the brain — BRAIN (The Brain Research Through Advancing Innovative Neurotechnologies).[4]

The key to understanding a complex system like the brain is to develop a new technology that is powerful enough, and has the complexity that allows us to understand the real, true significance of all the activity patterns. That is going to require computers to clock and analyse the data.

While we cannot imagine the future, we can be sure that such research will continue to accelerate technological developments, and impact us in ways that we cannot foresee.

References

1. Kale, Wilford. "Tourist Visits to State Sites Up 0.1% IN '90 — Blue Ridge Highlands Region has Biggest Gain, 6.1 Percent." *Richmond Times-Dispatch* (Richmond, VA), January 20, 1991. NewsBank Access World News.

2. Hubel, D. H. and Wiesel, T. N. "Receptive Fields, Binocular Interaction and Functional Architecture in the Cat's Visual Cortex." *The Journal of Physiology* 160, no. 1 (1962): 106–154.

3. "The DARPA Grand Challenge 2005." Defense Advanced Research Projects Agency. Accessed February 14, 2018. http://archive.darpa.mil/grandchallenge05/gcorg/index.html.

4. The White House. "The BRAIN Initiative." Accessed February 3, 2018. https://obamawhitehouse.archives.gov/node/300741.

5. How Algorithms are Altering Our Understanding of Systems

W. Brian Arthur

I want to talk about the difference that digitization, algorithms, and computation is making to science. We now have powerful new instruments, new technologies, that enable us to see what happened at the beginning of the cosmos, what is happening in the human brain or in animal brains, or deep in the oceans.

We are getting scientific technologies that did not exist 20, 30 years ago. We are seeing new phenomena. We are seeing completely new worlds we barely knew existed. I think this is going to be a golden era for science. I think we will look back on this era, and it will be possibly like the 1600s when we began to see so much.

The thing that fascinates me is that computation or digitization is altering the way we understand systems. Think of systems of highly interconnected setups whatever they are: entities, the immune system, the traffic system, the galactic system.

100 years ago, if we wanted to understand how systems evolved or changed over time, we had to do it with pencil and paper. It was pretty slow. You could get a set of equations. You work things out. If you were a superb mathematician like Poincaré, you could get some insights. Now all of us can take a system, put it on the computer, and do computer experiments, and watch how that system plays out on the screen. We are not so much looking for what is going to happen in the future. Rather, we are looking for phenomena. We are looking for patterns. We are looking for changes, things we would not expect.

I have been doing this. In the early days of my career, I started off with pencil and paper. But from the 1980s, all of us began using computation. And there have been quite a few surprising things that we saw. Quite unexpectedly, we saw far more threshold effects than we had expected. The system might look okay for a long while, and then suddenly 'bump!' It might enter quite a different phase or something very different, like going along in a dense traffic on the highway, and suddenly there is a traffic jam. You did not expect it. We are seeing this all over the place, threshold effects. We are seeing what you would call phase shifts. We might re-run the experiments, change the parameters a little bit. Nothing changes. Then we make one tiny tweak, and suddenly the whole regime is completely different.

These are giving us profound insights into our own world. Once we are able to use algorithms to look at our world unfolding or evolving, we begin to see that the worlds we have been looking at are organic. Conway's Game of Life is a little computer algorithm that looks very life-like.[1] If you look at some other system, it may look like a river basin getting carved out. We are seeing things that are unexpectedly life-like and organic.

For me, the biggest surprise of all that came up with computation, was that we could start with relatively simple rules. I am thinking of cellular automata here. You do not need to know what those are, just think of it as a very simple system that unfolds over time. You could give different types of rules. For some rules, you could predict what is going to happen in the thousand steps or million steps. But for nearly all the rules you can cook up, there is no way to predict, other than to allow the system to compute its way and see what happens.

This is a major shocker. This is not a casual result saying there is imprecision built into the computer or that you are not sure if you are entering the correct numbers. These are theorems in mathematics that are extraordinarily powerful. For most systems that are nonlinear, e.g. some system or rules at random, chances are the way they work out is no more predictable than waiting and seeing.

I think that computation is causing a very deep and scarcely noticed transition in science itself. And the way we think about the sciences — from neuroscience, biological science, physical sciences, cosmology, to economics — they are based less and less on equations, and more and more on algorithms.

Let me explain a little bit by what I mean. At the very start of the 1600s, people around who were doing pretty good science thought geometrically. Kepler worked geometrically. Galileo worked geometrically. Yet, there was a slow transition throughout that century. It was not until later in the 1600s, that more and more of science began to be based on equations.

Late in the century, in 1687, Newton did his theory of planetary motions, the *Principia*.[2] He did it with equations, and he cooked up calculus for that purpose. Then he went back so he could make it look scientific by expressing it all in geometry. Now something very similar happened in the 1900s. Alan Turing came along with the most influential paper in mathematics in 1937, showing something called algorithms, step-by-step instructions for a system to go through and calculate itself out.[3]

After Newton, we had equational systems. We had something on the left-hand side that told us how to update the system into equations. The equations were on the right-hand side. Those equations applied, and everything would work its way out. You could add some probabilistic terms later if you wanted. What Turing introduced was using equations in an algorithm. But those equations would describe a system that was changing, and that system would react to its internal changes, e.g. with if-then rules. So, if something happened, then you could use a different set of equations, or something different might happen.

This led to a world of equations, the base of any algorithm working their way out, but reconfiguring, changing, and adapting themselves to the conditions that they had created. That is the world we are moving into. It opens a much wider world where we have the benefits of geometry and equations, but we have the benefit of equations that are reacting to the circumstances they themselves have brought about. There are profound implications for this, philosophical ones, but there are also profound connections with biology. The biological system is a system that unfolds but reacts to its own unfolding.

When I was small, we used to think of physics as being the base classical science. Everything had to be like physics. That is changing. And we are looking much more to biology to inspire us. And computer science, that can change again. My point is that we are starting to think much more biologically. We are thinking algorithmically, and everything is starting to be based on information.

There is a huge subject I am fascinated by, called Algorithmic Information Theory.[4] It asks: what is the science of describing systems when it is information that counts? So, to put some of this in slogan terms: biology is basically the steady transformation of information.

References

1. Gardner, Martin. "Mathematical Games." *Scientific American* 223, no. 4 (October 1970): 120–123.
2. Newton, Isaac. *The Principia*. Amherst, NY: Prometheus Books, 1995.
3. Turing, A. M. "On Computable Numbers, with an Application to the Entscheidungsproblem." *Proceedings of The London Mathematical Society* 2, 42, no. 1 (1937): 230–265.
4. Chaitin, G. J. "Algorithmic Information Theory." *IBM Journal of Research and Development* 21, no. 4 (July 1977): 350–359.

6. How a Deeper Understanding of the Brain Might Solve the Mind-Body Problem

Terrence J. Sejnowski

There are many grand challenges in science, but which is the grandest? Us!

And perhaps the grandest question is one that was asked centuries ago, and encapsulated by René Descartes: "I think, therefore I am. *Cogito ergo sum.*"[a]

That created the mind-body problem. How is it that we can reconcile the fact that we have this body and this brain, and at the other end, have a conscious awareness of the world which seems to be disembodied? How could that be? Maybe once we've understood more about the brain, we will understand a little bit more about how it is that our internal world connects with the body and the external world.

What about us? What is the most exciting science right now? Every science has its era. What is happening now is an explosion of information. Amongst the sciences that are at the leading edge, is computer science. Biology is not just about energy, it is about information, and computer scientists know a lot more about information than biologists. This has really unfolded over the last several decades.

Another area that I am going to be talking about is the brain. It is an organ whose function is to get us through the world, to survive using

---

[a] Ref. 1, p. 5

information from the world and generating information. The brain is an information organ.

Finally, genomics is exploding, in terms of being able to sequence the gene of every single species, and in fact, every individual. Eventually, we will get sequenced. That will be a tremendous trove of information, not just about us but about the past too, because the history of evolution is written in the genes.

Now, I want to give you one concrete example of what I have in mind. It has to do with going from where we were just about a decade ago: from being able to record from one neuron at a time, to being able to record from all the neurons at the same time. This sounded like science fiction to me just a few years ago.

Let me illustrate how difficult it was back in the last century, recording from one neuron at a time. Suppose you can only see the world through a straw, one pixel at a time. Could you look out at a scene, and figure out what is in it? Well, I can look at your little eyebrow, and I can look at the little feature on your shirt, but to figure out how all that is connected together would be hopeless. There are just so many features, and to figure out how they all fit together would be a very difficult problem. However, if you could get a global picture by looking at the entire pattern, that might give you a clue.

Let's look at a study of neuron activity done on zebrafish.[2] We have a zebrafish. It is a larva, which has around 80,000 neurons in its brain. We have embedded this zebrafish in agar so it cannot move. There is no sensory input. In this case, when you have shut off all sensory inputs, there is no motor output. We can observe patterns changing in the brain, patterns that come and go throughout the entire brain. This is internally-generated activity, what we call "spontaneous activity".

What is going on in the brain? We have no idea! We are seeing here, for the first time, what we are up against. We are trying to interpret all these activities that are being internally generated, that are not connected to the world.

Most of neuroscience now is based on sensorimotor experiments. We flash a stimulus, we ask the monkey to respond. That is a different state from that of the zebrafish in the above-mentioned study. Spontaneous activity is a state that happens when you are sitting or lying on your bed, and your brain is just zipping away. You are thinking about what you are going to do that day, or what the problem is that is on the top of your head. Things just pop into your head. That is what is happening with the zebrafish, things are popping into the zebrafish's head: Oh my god, I can't move. What's going on here? How do I get out of this?

Well, I mean, I don't know what the zebrafish is thinking, but the point is we have reached the point where we can actually look at thought happening. And not just in zebrafish, but eventually in humans.

Feynman once said, "What I cannot create, I do not understand."[3] This is true of physics. You can't be sure that physics will take you to the moon until you actually get there. I'm not sure anybody doubted that celestial mechanics actually work, but here is a little fact of it that you may not be aware of. The astronomical unit back in the 1960s was off, and they did some radar bounces off on Venus in order to get sufficiently accurate numbers, so that when they send satellite probes out there, they could actually get the satellite there, and not lose it in space. You've got to get the numbers right.

As you learn more about the brain, you can embody it in technology. You can first simulate it, because that is what we do just to make sure that it actually works, but on some vastly slower timescale. Today, we can build chips that have similar algorithms embodied in them. What we have right now are massive clusters of computers that simulate just a tiny little piece of the brain.

However, if you want something to work and interact with us in real time, with all that computational power, you are going to have to have a technology that is basically as efficient as biology is right now. If you can actually reconstruct the neural circuits into silicone, you can vastly reduce the size, the power, and potentially scale it up. That technology is called neuromorphic engineering, it was started by Carver Mead.

The beauty of the brain is that it is asynchronous. You have information flowing through the neural circuit the way water flows, but there are nonlinear decisions being made at every stage. So that is a different style of computing in which you don't have to write down the program. Instead, you write down the network topology and the physiology of how the neurons respond. It is an embodiment of the program.

I think that as we learn more about the brain, we are going to have an added dimension to the way we understand ourselves. As we learn about what is happening in the brain, we are going to be able to build neuromorphic devices that are going to change how we interact with the world.

References

1. Miller, Valentine Rodger and Miller, Reese P. *René Descartes: Principles of Philosophy: Translated, with Explanatory Notes.* Dordrecht: Kluwer Academic Publishers, 1982.
2. Trapani, Josef G. and Nicolson, Teresa. "Mechanism of Spontaneous Activity in Afferent Neurons of the Zebrafish Lateral-Line Organ." *Journal of Neuroscience* 31, no. 5 (February 2011): 1614–1623.
3. *Richard Feynman's Blackboard at Time of his Death.* 1988. Image Archive, California Institute of Technology, California. Accessed Jan 2018. http://archives-dc.library.caltech.edu/islandora/object/ct1%3A483.

7. The Smallest Insect is More Complex than a Star or Galaxy

Martin J. Rees

What I am going to do is to say a bit about my own subject, astronomy, and the limits of science in general.

We would like to understand a bit more about galaxies. If we are particle physicists, we would crash particles together in the lab, but we cannot do that with galaxies. We can though, in the virtual world of a computer. We can create a simulation showing two galaxies crashing together and merging a real sort of train wreck there (I should warn you that in 4 billion years, the Andromeda galaxy is going to crash into our Milky Way).

We have inferred by looking at distant galaxies and by other observations of radiation, etc., that our universe started off in a hot dense state 13.8 billion years ago. That number is known to 1% accuracy.

Why is the universe expanding the way it is? Why does it contain atoms and radiation? To answer such questions, we need to go back much further. We do not understand the physics that well, since it is beyond the range where we can check experimentally. But we are working on it.

One exciting idea favored by many theorists, is that our Big Bang, if we understood it properly, would prove to be not the only one. It may be just part of some vast multiverse. This is speculative. But in 50 years' time, I hope that we would have firmed up the evidence for a multiverse, just as how we did not know what the Big Bang was 50 years ago. Now we are able to talk seriously about the Big Bang, back to a nanosecond and say things about it still earlier.

I think it is very interesting that there are lots of links between the micro world and the cosmos. The everyday world of people and mountains is determined by atoms and molecules, and they stick together. Stars are fueled by nuclear reactions within the atoms. There is a link there, and there are other links. In the micro world, the dominant physics is quantum theory. In the cosmos, Einstein's theory holds sway.

Now, general relativity and quantum theory are the two twin pillars of 20th century physics. But as you probably know, they have not yet been meshed together. A unified theory of the two is an unfinished business. Now, in most contexts this does not matter. But if we really want to understand the beginning of the universe — not just the first nanosecond, but the first tiny, tiny fraction of a second, where the conditions were imprinted — we have got to achieve the synthesis between quantum theory and gravity, because right at the beginning, the whole universe was squeezed into a microscopic size. Quantum fluctuations can shake the cosmos. And in order to have a theory that can understand that, we need to have this unification, which eludes us; it is still unfinished business. This unified theory may tell us whether there is a multiverse. That would be a surprise, but it will be a further Copernican revolution as it were.

Our galaxy is one of billions of galaxies accessible to our telescopes, but its entire panorama may be the aftermath of one Big Bang among billions of big bangs. That is speculative science, but it is an exciting prospect that widens our horizons in an almost mind-blowing way.

There are two frontiers of physics: the very small, on one hand, and the very large on the other. Yet, 99% of scientists work neither here nor there. They work in the third frontier, which involves very complex insects, peoples, and mountains. There are things large enough to have layer upon layer of intricate structures, but not so large that they are crushed by our planet's gravity. The smallest insect is more complex than a star or a galaxy.

This is the biggest challenge of science. That is why it is not surprising that we understand stars and galaxies quite well, but our understanding of familiar matters of interests, be it diets or childcare, for instance, is still so meager that expert advice changes from year to year. It is ironic that you should not believe anyone on those subjects whereas I hope you half-believe me when I talk about stars and galaxies.

The sciences are sometimes likened to a hierarchy, different levels of a tall building. Math and particle physics at the bottom, then the rest of physics, then chemistry, then cell biology and so on up to psychology, and then the economist in the penthouse.

There is a corresponding hierarchy of complexity: atoms, molecules, cells, organisms, societies and so forth. This analogy with a building is a poor analogy. In a building, insecure foundations imperil the rest. But the higher-level sciences dealing with complex systems are not imperiled by an insecure base. Each science has its own distinct concepts and explanations. For instance, mathematicians trying to understand why flows go turbulent or why waves break, do not care that water is H_2O. They treat the fluid as a continuum. And everything however complicated — breaking waves, migrating birds, tropical forests — is made of atoms and it obeys the basic equations of quantum physics. But even if those equations could be solved for immense aggregates of atoms, they would not offer the enlightenment that the scientists seek. Reductionism is true in a sense but not in any useful sense. Each science has its autonomous concepts and laws, each is equally fundamental.

Obviously, some topics get cleaned up and codified (atomic physics for instance), and researchers then move on. But maybe we should be open-minded about the obverse possibility that we may eventually hit the buffers because our brains do not have enough conceptual grasp. Einstein averred that the most incomprehensible thing about the universe is that it is comprehensible.[a] He was right to be astonished because our minds have not changed much since our ancestors roamed the African savannah and it is amazing that these minds can comprehend so much of the counterintuitive world of atoms and the cosmos.

Nonetheless, and here I am sticking my neck out a bit, maybe some aspects of reality are intrinsically beyond us in that their comprehension

---

[a] Ref. 1, p. 24

would require some post-human intellect, just as quantum theory is beyond the brainpower of monkeys. This links to the subject of computers and artificial intelligence (AI). The computers will understand and be able to make predictions but that does not necessarily mean that we will be able to grasp these complications. We will hit the limits of what humans can really grasp even if there are no limits in principle to what we can compute.

Experts disagree on how long it will take before machines achieve general-purpose human level intelligence. Some say 25 years, others say never, and the best average in a recent survey was about 50 years. But some do claim that once a threshold is crossed, there will be an intelligence explosion. That is because electronics is a million times faster than the transmission of signals in the brain, and because computers can network and exchange information much faster than we can by speaking. And AI will perhaps eventually make the great discoveries, using reasoning that we, humans, may never be able to understand.

Simple life may be common and intelligent life may be rare. Yet, these thoughts incidentally lead me to believe that if we ever detect artificial signals from intelligent life in space, these signals may come, not from any kind of organic civilization, but from machines. No one is holding their breath for that, but I am actually involved in some research to improve the chance of detecting any artificial signals out there.

Well these predictions about AI are exciting, but they are rather scary. Just as scary are some of the predictions about the use of biotech, which has been advancing equally fast.

In fact, twelve years ago I wrote a book on this theme, I called, *Our Final Century?* My publishers cut out the question mark and the American publishers changed the title to *Our Final Hour.* Americans like instant gratification. The book addresses two concerns: First, about humanity's ever-heavier footprint on the biosphere, but also about the risks for empowerment of individuals and small groups by these powerful technologies. I think we do need to worry because as I put it:

"'Spaceship Earth' is hurtling through the void. Its passengers are anxious and fractious. Their life support system is vulnerable to disruption and breakdowns. But there is too little planning, too little horizon-scanning, too little awareness of long-term risks."[2]

We will have a bumpy ride to this century, but we should remain upbeat. We should not put the brakes on progress. We should have to redirect it and realize that the precautionary principle has a manifest downside. We need more technology but guided by the social sciences and ethics.

I shall give the final word to the great scientist who, I was sadly too young to know, the great biologist, Peter Medawar. I quote:

"The bells that toll for mankind are [...] like the bells of Alpine cattle. They are attached to our own necks, and it must be our fault if they do not make a tuneful and melodious sound."[3]

References

1. Vallentin, Antonina. *Einstein, A Biography*. London: Weidenfeld & Nicolson, 1954.
2. Rees, Martin. "Surviving the Century." Lecture at The Inaugural Carl Sagan Lecture, Cornell University, May 8, 2017. http://www.cornell.edu/video/lord-martin-rees-carl-sagan-lecture.
3. Medawar, Peter. "The Future of Man." Lecture at The Reith Lectures, BBC Radio 4, December 20, 1959. http://www.bbc.co.uk/programmes/p00hg13w.

8. The Past, Present, and Future of the Biosphere

Eörs Szathmáry

The great challenge to our biosphere seems completely unrelated to the biosphere itself. Issues in society, economics, and technology may seem irrelevant, yet they contribute to species extinction and an increase in ecological footprints. This is something we need to address. It is not in school textbooks, nor is it discussed in schools. And by the time people are informed about it, it might be too late.

Does the past help us better understand how we came about? Does the present help us to better understand the limits of sustainability from a biological point of view? And does our envisioning of the future lead us to maintain a stable ecosphere? What complications might we encounter along the way? It is important that we examine these complications as these are what the younger generations will need to deal with in the future.

First, let us consider the past. The past of the biosphere, is evolution. It is more than 3.5 billion years old, and it has been going on for quite a while. This is an incredibly long time. There are two major challenges to our scientific understanding.

The first major challenge concerns the origin of life. It is a notoriously difficult problem, for the following reason: we do not have a convincing scenario for the origin of life, and we do not know whether life is likely, very likely, or moderately likely on another planet like the Earth. We know even less about whether simple life can begin under radically different chemical or geological environments, and what that probability is.

You could ask the question: "What is the chance that a chemistry on a planet will produce systems suitable for the generation of life?" But chemistry is not developed enough to answer this question. In fact, chemistry cannot even answer the question for the conditions that prevail on Earth, let alone for conditions that are very different from Earth.

The scary thing is that it might actually very well be the case that we are the only intelligent life in the universe. This means that the value of the biosphere is much greater, and in that case, our responsibility for it is much greater.

The second major challenge concerns the origin of intelligent life. Whether or not there is intelligent life out there, this is something that we do not know. It is very unlikely. But the probability of the independent origins of intelligent life, in my expectation, is much smaller than for life itself, which might actually be small enough, so this is a conditional probability kind of thing.

John Maynard Smith and I have a list of what we call, the major transitions in evolution.[1] It starts with replicating molecules, and it ends up with human societies with language. You could ask the following question: "What fraction of planets would evolution progress along somewhat similar lines as us, once their lifeforms are able to self-replicate?"

Well, to tell you the truth, we don't even have a clue! I have a hunch that if there is life on other planets, most of them will be something like bacteria. We do not know whether there is intelligent life beyond our biosphere. We do know that it is very unlikely. At the same time, we simply do not know what kind of intelligences might exist beyond Earth. That is, again, a very interesting problem. At the moment, we can only extrapolate from one experiment, that is from our biosphere.

Evolution theory is at the centre of the biological sciences. But it is not a frozen science. Our knowledge and insights on evolution will deepen. I predict that in the next 25 years, there will be a lot of developments, and we might soon be able to find answers to questions about evolution: What makes a certain lineage more or less evolvable? That means, will it respond to directional selection faster or slower? What kinds of things can speed up or slow down evolution? Perhaps in the future, our deeper knowledge of evolution may shed light on whether we humans are alone in the universe or not.

But what if we are indeed alone in the universe, then this biosphere is very precious. But preserving our biosphere is not an easy task. Allow me to highlight a few challenges that our biosphere is facing at the moment.

The rate of background extinction of species in the biota is a few species per year. This happens for natural reasons. However, there are human-induced extinction of species, and that happens at a much faster rate, to the extent of it being 1000 or even 10,000 times more than the background rate of extinction. If it continues like that, in the year 2050, we might lose 50% of the existing species of the world. That is a man-made mass extinction, and we have to do something about it.

We also have a problem involving the ecological footprint, which refers to how much land and sea you are using per year as a kind of natural capita to sustain life. The main problem is that the big ecological footprint of civilization is now 1.5 times the existing natural capita of the Earth. The increasing human population is increasing the rate of species extinction. It is increasing the ecological footprint. Ultimately, we are going to get into a self-reinforcing spiral, a positive feedback. That is not going to work out in the long run, and we need to figure out how to put a stop to this positive feedback spiral.

There are a number of things that we need to concentrate on: how to deal with the problem of species extinction, and the preservation of the biosphere. These link to many problems related to the human population, and how we might deal with them in the near future. For example, what kind of population numbers will be sustainable? What kind of economic policies will be best suited for preserving the biosphere?

You could argue that it depends on many things. If everybody just eats rice, you can have up to 20 billion people on this planet. But does everybody want to just eat rice? That is quite unlikely. Attitudes are a difficult thing to change. And we have many attitudes that must be changed for the sake of preserving our biosphere in the future.

The traditional way of farming meat is not sustainable. It is very inefficient to rear cattle (not to mention the fact that cattle are producing a lot of methane, which is a greenhouse gas). Unless we are willing to cut down on the production of meat, or change the way we produce meat, there is very little that can be done. We need a change of attitude before we can make any further headroom.

There are economists who are always talking about economic growth. But such growth is done at the expense of the increasing ecological footprints and the declining natural capital. Such growth is incompatible with our biosphere. And I am glad that an increasing number of economists are now saying that you cannot have economic growth at the expense of a declining biota. Economic growth has to be limited to those cases where the biota is still preserved.

That is excellent! But is there a problem?

The problem, I'm afraid, lies largely at the political level. The problem with democracy is that politicians tend to plan on a very small timescale, typically four years (the period between elections). But our biosphere does not operate on such a timescale, and we cannot afford to plan and operate on a four-year timescale. Something needs to be done. We need to advise our politicians to understand the situation and go beyond their usual four-year plans. Their considerations should be more than just about increasing GDP.

There is a proverbial case of a poor country with a lot of forests. The people cut down all the forests, and the next year, you see that the GDP has gone up. But what happens after that when there are no more trees?

Reference

1. Maynard Smith, John and Szathmáry, Eörs. *The Major Transitions in Evolution.* Oxford: Oxford University Press, 1995.

9. The Public Engagement of Science and Society

Helga Nowotny

You have heard fascinating stories where science is working at the frontier of knowledge: of what we know today, and what we think we will know in the coming century. Science is not isolated from society, and I want to look at the challenges that arise at the interface between science and society.

What you have heard depends, first of all, on the place, the attitudes, the kind of appreciation that science has in society. If we look around the world today, we see that science is not evenly distributed. There are leading countries, hotspots, places, or universities that you want to go to, places where the working conditions are excellent. And there are other places that look more or less like a scientific desert. This phenomenon is partly the result of history (of course, everything is partly the result of history), but partly, it has to do with the appreciation that a society gives to science. What you have seen is funding that has been received. We have universities, we have research institutions, we have the prospects for future careers. All this must be funded from somewhere, and the kind of science you have been listening to needs public funding. Let's be clear about that. Private funding is always welcome, but the kind of frontier research comes from public funds.

Way back, in the 1930s, there was an American by the name of Abraham Flexner. He wrote a manifesto with the title, *The Usefulness of Useless Knowledge*.[1] This manifesto was a plea to set up an institute to fund research where you do not know where such research will lead to.

A couple of years later, he was successful in getting the funding he needed. In this case, being in the U.S., it came from private money, but it led to the setting up of the Institute of Advanced Study in Princeton.

In the beginning, the first people to be employed in the Institute of Advanced Study were mathematicians, historians, and a few others. The institute was dedicated to the usefulness of useless knowledge. Alan Turing was there for a couple of years. John von Neumann bu ilt the first computers in the institute. In fact, the institute became the place in which computational science was born, and that made a tremendous and useful contribution to the U.S. at a time where it was entering the Second World War.

This is one of the first challenges: We need to remind politicians and the general public of the usefulness of seemingly useless knowledge.

Politicians, as we know, want applications and deliverables. They want to have results before you have begun your research. In some ways, this is okay. This is what politicians have to insist on, because it's taxpayers' money. Nevertheless, we have to make a special effort — and it is a continuous effort, I can assure you — of convincing them that for the 21st century to continue, it is essential that investment goes into this kind of frontier research, where we do not yet know what the outcome will be.

This journey into the unknown is a kind of research that is inherently uncertain. You do not know where it will lead you, you do not know the outcome. What we know from history (and this has been confirmed many times) is that it takes about 50 years from the first breakthrough to arrive at a broad range of applications. We also know this from medication and drugs. It takes about 20 years to come up with something useful, something that can be put on the market, and so on. There is uncertainty. You do not know where you will actually get what you will get, and when you will get it. This is one of the largest challenges vis-à-vis the funders that I see.

We also know that there have been some moving testimonies to this. For instance, François Jacob, the French Nobel laureate, spoke about day science and night science.[a] Day science is the glorious side of science: the glossy brochures that every university puts out to show what is being done, the product fairs, etc.

Then there is night science, and night science comes with a lot of uncertainty and some frustration. You set up an experiment, it does not work. You have to go back and ask yourself, what did I do wrong? And you start again.

These are the two sides of the same coin. Just as day and night belong together, so are these two sides of science. We take far too little effort in convincing the public and society of the usefulness of useless knowledge, of the kind of inherently uncertain research that we want to pursue. Instead, far too much emphasis has been made on day science, impressing the public with the finished results.

What gets lost in this engagement with the public is the process of how scientists got there. It is this kind of engagement and participation with the public, that gives them a feeling of what the process of research consists of. We have to let them catch a glimpse of night science as well. We have to let them in to see the process of research, and not just the finished results. Let them partake in the fascination, this wonderful excitement that comes when you are on to something. You don't know. It may not materialize. You start again. But you have this feeling that you're on to something.

You have also heard about the importance of technology. Indeed, technology is very important, but it is not exactly the same as science, nor is it the same as innovation. Very often, we mix up these concepts. Technology is much older than science, but historians of technology and historians of the economy tell us that somewhere in the second half of the 19th century, there was a switch to what is called science-based technology. In other words, there was a new interaction that came about, at that time, between science and technology. Science-based technology means that the technologies we use are an embodiment of scientific knowledge, in a sense. At the same time, we can also see how important

---

[a] Ref. 2, p. 126

it is, the kind of questions that are opened up by using a new technology. Fed back into science, they allow you to pose new kinds of questions that you could never have posed, knowing that you have no tool to answer those questions then. So, this very close interaction is important.

Let's come back to what politicians and society expect from what we now often call the techno-sciences. Seen from the outside, we do not see how exactly technology and science fit into each other, but we now see that something is going on where they are entangled. This leads to the kind of technologies that society is familiar with. Think of it as the end of a process. You are the generation that grew up with having a certain kind of technology in your pocket, and you can see how your children are already discovering what you can do with this technology much earlier than you were able to do it, and so it goes on, and on.

This will change the workings of our economy and society. In the 1930s, there was a student club in Madrid, and at that time, the members made a point to invite some of the most illustrious speakers of science from around the world. They had invited Albert Einstein, but they had also invited an economist by the name of John Maynard Keynes. These were the years of the Depression, both in the U.S. and in Europe.

Keynes chose an unexpected topic, he chose to speak about the future for our grandchildren. He made a big leap forward, and he said, "Let's look at the year 2030, a hundred years from now, and see what the world will look like." By and large, his predictions were limited to what he called, "the mature economy."[b] Basically, he looked only at the U.S. and Europe. By and large, his predictions were correct: people will be better off, etc.

But then he made a prediction that leaves us a bit stunned today. He said that in the year 2030, we will, on average, work 3 hours per day. Think about it, we have not arrived at 2030 yet, but with the automation industry proceeding very rapidly, you see that we will have a problem. By the way, for him, it was a moral problem: what will people do with their time?

---

[b] Ref. 3, pp. 358–373

For us, it is a political problem, it is an economic problem, and it is an unresolved problem. What people see are the technologies that change their life, and if people are out of work because everything becomes automated, this is not a very good situation to get the kind of support from society that you want, and the support that you need. This means that something needs to be done on that part of the interface: listening to what people are worried about, and their perception of what science and technology will bring to them. At the end of the day, people ask questions: What are you doing for us? For me? My job? My family?

A lot of research has been done on risk and risk perception because scientists have noticed that not everything that science or technology produces is welcomed with open arms. And people have different risk perceptions. I have been struck by how science demonstrates its ease and willingness to embrace uncertainty. Science thrives on this cusp of uncertainty. Society — and we are part of society — likes to have certainty. Sometimes, there is even a craving for certainty. We want to know what is going to happen. There is also this misconception that we can turn to scientists and ask them, "Tell us, what will happen?", without realising that very seldom are scientists able to give a clear yes or no. They will tell you "Yes, under the conditions of...", "No, under the conditions of...", and they are right, this is the only correct answer.

Now, this is not what politicians want to hear, and this is not what most of us as citizens want to hear. We want to have this certainty. So, there is again something to be done, a challenge in how we convey what uncertainty actually means and what those risks are. We also know that most people worry about the wrong things. There have been many, many studies comparing the actual occurrence of events that are to be avoided due to fear, and the risk perception that people have.

What people don't worry about are the risks that they do not see and are not familiar with, namely, systemic risks. These are things you do not see. They build up, you do not see them coming, then all of a sudden, they are here. And they hit you. Again, this is something that needs to be conveyed to society, that these systemic risks exist.

I want to end with a new opening of science that I see happening, one that has already started at the beginning of the century. In fact, there are always antecedents. The astronomers always had a lot of support from amateurs who were fascinated by what astronomy did. Botanists in the 19th century relied on amateurs ('amateur' means 'lover').

But in this century, what we have seen is something called "Citizen Science". Now we're seeing a revival of this, and it takes on different shapes and forms. Yet, it is a kind of movement that comes partly from the fact that we are the most educated societies on Earth today that has ever existed. This is something we must remember. The formal education that people get (and the access to knowledge and information) open new doors.

Citizen Science uses these open doors, but an open door is always open towards b oth sides. Scientists are welcoming the open door. They have set up ways of using, inviting, and accompanying laypeople who want to participate in the process of research.

This has happened some time ago, in biomedicine where patient groups have shown to have much more knowledge because they lived with their relatives or family members who have certain conditions. They have shared their knowledge with doctors and people from the life sciences.

We also see people who are computer-savvy, who work on their laptops, for instance, to help a crowdsourcing problem, such as how a certain protein is folding, etc. In the environmental area, we have lots of people who are using apps and other technologies that are widely available today to help monitor oceans, and so on. We know relatively little about who these people are, what motivates them, what their expectations are, or how we can sustain this process.

There is an ongoing open door here, a movement of Citizen Science. It is not fully developed yet. Volunteers come and go, so we also need to become more knowledgeable about what is happening there and avoid exploiting them. Once people start to feel exploited, they will drop out. We need to know how we can sustain their participation.

Nonetheless, I think bringing them into the research process itself is the greatest benefit. We need more volunteers in the life sciences, people willing to give their genome and tissues for public research. Much of these is needed, this opening of science to keep the momentum that is being created is very important to make people feel like they are a part of this exciting venture.

I think there is an important distinction to be made regarding contributions to science. Harry Collins was a social scientist studying people who were investigating the gravitational waves for many, many years. He spent a lot of time with these people and was able to get to the point of understanding them speaking their language (of science). He even made a sort of Turing test with them, and he passed the test as a gravitational physicist.

Collins made an important distinction: there is interactional expertise, and there is contributory expertise.[c] For interactional expertise, you can reach a level where you can talk to one another, but you won't be able to contribute to the scientific advance. I think this is a useful distinction, so as not to overstate the expectations that you have but also not to put people down either.

Science is the best invention that humanity has made to bring the future with all its enormous potential into the present. But science needs society and the close interaction with society to shape it. Tapping on the potential of Citizen Science is one way of tackling what I see to be the greatest challenge of science.

References

1. Flexner, Abraham. "The Usefulness of Useless Knowledge." *Harpers* 179 (July/November 1939).
2. Jacob, François. *Of Flies, Mice and Men*. Cambridge, Massachusetts: Harvard University Press, 1998.

---

[c] Ref. 4, p. 127

3. Keynes, John Maynard. "Economic Possibilities for our Grandchildren (1930)." *In Essays in Persuasion*. New York: Harcourt Brace, 1932.

4. Collins, Harry. "Interactional Expertise as a Third Kind of Knowledge." *Phenomenology and the Cognitive Sciences* 3, no. 2 (2004): 125–143.

Part II

Part II

10. Artificial or Augmented Intelligence? The Ethical and Societal Implications

Eörs Szathmáry, Martin J. Rees, Terrence J. Sejnowski, Tor Nørretranders & W. Brian Arthur

Terrence J. Sejnowski: Back in the 1990s, IBM had developed a computer hardware called *Deep Blue*, and this was a brute force project. That is to say, when you're playing chess, you have to look forward. If I move my knight to this location, how can my opponent counter that? The further you can go into the future, the more accurate you can be in terms of picking the best move. So they created a computer that simply could outpace a human in terms of how many plies you can look into the future.

Now that is not the way humans work. Humans play chess with pattern recognition. An expert chess player will look at the board and he will know immediately which two or three possible moves are most likely to be the best. That is because he has played thousands and thousands of chess games and seen similar board positions. He'll take to ten plies. But that is as far as the human will go, whereas the machine could take it to twenty plies.

It had an interesting impact. A lot of people thought that if Kasparov, the world's best chess player, cannot beat the machine then it will be all over for humans. Basically, the chess machines are going to play each other and humans are going to watch.

That's not how it played out in the end. And it was very interesting. What happened was that because these chess programs became generally available, and since personal computers became as powerful as the *Deep Blue*, everybody had a chess program that could play at the grandmaster level on their laptops. That means that it could help improve your game in the same way if you want to learn how to play tennis. You are better off playing somebody better than you because that will elevate your game. Now it has reached the point where the level of human play has improved over the last couple of decades.

Interestingly, it also helped advertise chess because it used to be the case that if you wanted to be a world-class chess player, you had to live in a big city that had a world-class chess club and that will allow you to play world-class chess players. But now, if anybody can have a world-class chess player in their laptop, it means it does not matter where you come from. In fact, Magnus Carlsen, who is now chess champion, grew up in a small Norwegian village and he has reached the highest levels. One of the lessons from this is that we should not think of AI as replacing humans but enhancing our abilities. There are really more of them right now. I would call them cognitive appliances, in the same way that we have airplanes and things that can help improve our ability to store information. This is just a more cognitive version of that at helping us improve our brains.

Eörs Szathmáry: May I raise a question? I fully agree with you that this is what we should use artificial intelligence for. Not to replace, but to augment human intelligence and thinking. But I think although that is an eminent goal, there will be an increasing number of approaches and dimensions where these replacements will nevertheless happen. That raises a few very important questions.

So imagine (I know this is a blunt example) that they put the *Encyclopedia Britannica* as a microchip inside your head. People might ask: why should I learn? Should I learn anything at all? If we are using this prosthetic more and more, there might be a decline in the biological power of the people because there will be more reliance on this kind of extra add-ons, right? If you take vitamins from fruits all the time then you can afford losing the genes for the synthesis of vitamins.

Now you might say that is not necessarily a disaster because there will be a more and more integrated type of life between humans and these artificially intelligent gadgets. That is fine until we reach a point of disaster. By then people are not going to be worried about manufacturing gadgets that they can put into their heads because they would be worried about food and water. As Einstein said, the next World War might be fought with sticks and stones.[1]

I think that is a very important reminder that although this kind of development about AI might be welcomed, it might lead to an increased dependence on these devices on a societal scale, which under circumstances might end up in an excellent disaster.

Terrence J. Sejnowski: Well, it has already happened. So for all practical purposes, the Internet has made me omniscient. I can ask any question on Wikipedia and within a few seconds, I get all the vast knowledge that is on the Internet. Now it is true it becomes a symbiosis. In other words, it does become something that I rely on.

But here's what I think. It is not like a chip that replaces it. The learning now is not learning of facts but learning how to access facts. In other words, in order to use Google or the Internet properly, you have to understand the protocols: which key words should you use, how to do things literally. That, in itself, becomes an expert talent or skill you can acquire. But it is symbiotic in the sense that the Internet is always changing and so does your brain as they both interact with each other.

But you are absolutely right that children are no longer taught by rote in the same way that we were, at least I was. When I went to college my computer was two sticks of wood; it is called the slide rule. One of the virtues of the slide rule is that you get the answer pretty quickly to two and a half significant figures. But you had to figure it out yourself, which means that you had to get used to thinking about magnitude of numbers. Now I teach, and I get exams back where the students are off by ten orders of magnitude. They do not have any sense for a reasonable number because they rely on the calculator to give them some number.

It is already happening. The calculator is an early example of that. But I think that, on the whole, it is going to enhance what we are doing. In fact, there is a group that claims AI should really be called Augmented Intelligence, rather than Artificial Intelligence. That is what I would propose as being a better route for the future.

W. Brian Arthur: If you look at this from the point of view of Minsky or John MacArthur's (in the 1950s and 1960s) that started all this and coined the term artificial intelligence, it has been incredibly slow. It has taken decades to build up to a decent AI program that will play very good chess. But from the point of view of how long human beings have been around, this is staggeringly fast. This has happened in the last 50 years.

I am interested in something that is called AI complete problems. It is a whole set of problems, where it has been very roughly held that if you could solve one of these problems you could solve any of them. These are problems like the ability to recognise objects, faces or animals; the ability to pick out something in a scene; the ability to translate languages, to make reasonable semantic sentences not just the words; the ability to read something and not just recognise the characters but get a sense really of what is going on. There are maybe half a dozen of these major problems and I do not think they have been solved by any means. But in the last five years or so we have gone up quite another step. Of course, that is due to deep learning, which is quite incredible.

I read a book written in 2010, which said that someday we might get driverless cars but no time soon. As for driverless trucks, it might happen, but not in our lifetime. Well they are here and the reason they are here is that they can recognise what is going on generally. They can look at a scene and make out there is a road. They can differentiate a poster of a small child from an actual small child, and slow down and watch for that. It turns out that the limitations to driverless trucks and transport are largely legal at the moment, having to do with insurance and liability. So the limitations are largely institutional.

We are suddenly entering a different era. I can walk through London airport and these AI programs recognise my face. The computers are trying to make an association. They see a set of pixels. They can associate that with me and my passport. I walk down to the departure gate. There is another station I have to stop at and get photographed again. It will recognize me moving for another two hundred yards. Computers are starting, in a sense, to think like humans, in a way that they did not do it before. So I think that something deep has happened.

In the 1960s or 1970s, we would have said that computers are going to do massive calculations. Their electronic brains are almost infinitely smarter doing arithmetic. But now what has happened is that they are kludgy, but they starting to be able to think a little bit like humans. Like most people at Silicon Valley, I find this stimulating, fascinating, and creepy all at once.

I think that there is a problem and we should at least address them in this group. What is the future for all of this? Where is it taking us, ethically, and as human beings?

Martin J. Rees: We expect that they will be replacing human drivers and things like that. But I think there could be a downside even in the short term because it might lead to a sort of descaling of the drivers. I know airline pilots worry about this in the sense that most of the flying is automated now. But if something goes wrong, then the real pilots will take over. And they made the takeover so well. It is thought that the occasion when a French airliner crashed in the South Atlantic was perhaps due to the fact that there was an emergency and the pilots have been sort of off-duty. They came to realise something serious had happened and they did not take the right action because they were not quite up to speed as to what was happening.

There is a downside in this transitional period where the computers cannot do everything. We have to worry about breakdowns, and cyber flaws of various kinds. But I think one thing that is clearly going to happen is the ability of computers to process huge amounts of data and seek small correlations just above the noise is going to increase. I guess this is what the corn hedge funds are doing already with their computers. That is how they profit from the rest of us by having this advantage, and that advantage is probably going to grow.

Eörs Szathmáry: I have a few remarks. One remark is that the reason why we are so good at general intelligence is biological. We have to complete a life cycle. So that means we have to find our way around society. We have to get our food, educate our children, and that is what I mean by the completion of the life cycle. So I predict that it might actually be very worthwhile for the artificial intelligence's development to actually try to simulate that, to raise robots rather than design them.

The second point I have to say that in terms of the yield, so to speak, we are still much smarter than any of the achievements that you can have in artificial intelligence. It is amazing that despite our biological limitations, we have achieved so much. Relative to what we have, there must be some very clever things in our brains that we do not yet understand.

I guess there will be a fruitful interplay between neuroscience and artificial intelligence and there will still be many surprising things to come. That is my expectation and prediction.

Terrence J. Sejnowski: I totally agree, and I think that we have reverse engineered maybe two algorithms from the brain. What I think you are referring to is something to do with global integration. Right now, for example, we have the equivalent of a visual system to recognise objects, and then we have the equivalent of a current network that does sequences of words. But what we do not have is a way to produce an integrated sort of global perception. So when I am looking at the world, sensory information is coming through my eyes, ears, body, and it is all integrated into a single sense of being present, of interacting. That is a mystery, and some people call it consciousness or conscious awareness. Francis Crick was going after the visual side of it. Some people want to go after the social aspects of the fact that I am aware of you as an autonomous agent and that is called the Theory of Mind in psychology.

Big data is king right now in Silicon Valley. It is very valuable and from it, you can extract profits. These are profits from advertising, profits from being to make predictions for people and what they would do in terms of voting.

A few years ago, I was at a Neural Information Processing Systems (NIPS) meeting, where deep learning arose from that community. Mark Zuckerberg, the founder of Facebook, showed up at one of our workshops. He was there to announce they had just started a new AI lab. Yann LeCun, who is a good friend of mine and was going to be the director, introduced me to him and we had this discussion about neuroscience. It turns out that he had been looking into the brain. He had asked me all these very intelligent questions and was particularly interested in the theory of mind. He also asked if I could send him some references, which I did.

Surprised that he would be interested in references, I talked to Yann afterwards that it is really quite remarkable that someone who is worth many billions of dollars would be interested in the brain. He said, "Terrence J. just think about it. Facebook has more data on more people in the world about what they like, who they know, and how they interact with each other. They are going to create a theory of every person's mind on the planet, so that they can know you better than you know yourself or your spouse knows you."

This is both fascinating and creepy. It really is reaching a point where at least in some limited domains, this is going to vastly overtake. I might interact with a few hundred people well enough to sort of predict what you are going to say or how someone is going to vote. But if you can do that for billions of people, just think of the power you have.

Martin J. Rees: I guess in a sense you were saying that we integrate information about this room via all channels. But in the case of the supercomputer, it can do the same integration over the whole world in a sense, insofar as it could be aware not just of any local region but aware of everything at the same time.

Terrence J. Sejnowski: You brought up the driverless car. Let's imagine a scenario. You are learning through experience. As you get new experiences, you can integrate that into your decisions. So imagine that somewhere in Australia, they introduce driverless cars and the program comes across an object that hops across the road. It has never seen anything like that. It treats it like a dog and manages to avoid it. But what it does is that now, because it is a new experience, it integrates it into its new learning process. Not only that but it spreads into all other cars in the world, so that something that happens locally can then be broadcasted to become part of the whole global knowledge network in all cars. I think what is going to happen is that it is going to be much more globally integrated in a network of interactions between many different cultures, societies and conditions.

Martin J. Rees: But some people worry that if this entity gets a mind of its own, we need to get rather scared, because obviously nothing matters when you actually compare it to the human brain, but eventually it will. There must be some advantage in the interconnectedness at the speed of light and the fact that signals go, I guess, a million times faster than they go in our heads. So, in the long run, one can imagine that the takeover by machines will be pretty complete. The question is how are we going to cope with that?

W. Brian Arthur: Well, AI really concerns me and especially what you have just said. At what stage will AI take over from human beings? I have thoughts about that and it goes back to what Eörs just said. So imagine that there are new entities that are very computational, being created all the time and those are called human babies. They come into the world and they need to come up to speak cognitively. So lately, meaning the last few hundred years, we have been putting them in school and we have been educating them. This is fairly new to human race. We have been educating them, giving them facts, teaching them how to reason, teaching bits of logic. In kindergarten, we teach them to share toys, behave, and this goes on for quite a few years. Modern cultures could go on for 10, 15, or 20 years. I like this idea that we should not just program computers. We should raise and educate them.

This has an ethical overtone that comes to your question. When we are bringing up human beings, the major part of education for the first few years is to teach children ethics. We want to make sure when children grow up they can deal with other human beings. We want them to have some social and emotional intelligence. It is not perfect, but I believe that if we are worried about computers or AI taking over the world, we are going to have to do something similar with AI machines. As we are raising these machines and teaching them with how to deal with other machines or teaching them how to deal with human beings, we need to inculcate ethics. There will probably be pirates, outliers, or outlaws. We can circumvent all that. We will have to go after those, so I would assume that there will be a class of fairly well-behaved machines and every so often we have to go out and assess what has gone out of control. I think that is what we are facing over the next hundred years or so. Ethics is going to have to be built in.

Martin J. Rees: I am slightly worried that the ethical code that people would get depends on where they are brought up. You can imagine that robots get very different ethical stances depending on where they are. But the other thing is that we humans are constrained in our bodies. We are individuals. As computers develop, it is not at all clear to me if they will behave as individuals. They will be networks and so they will be more like a single mind, than like a society of individuals. It would be inappropriate to think of them as a civilization of robots, especially if they think in a sort of collective mind, like a hive of bees, or something like that.

Also, we got to bear in mind that we have to build in basic ethics and common sense in the early stages. The classic example is if the Internet in your home is supposed to keep your fridge full, it may put the cat in the oven because that is the only way to get meat. But the AI will not realize that it is inappropriate to do such a thing.

I think that is the sort of problem that will happen quite soon.

But in the long run, I think the key question is to what extent will they be a single mind.

Terrence J. Sejnowski: I think we have to be a little careful. Humans suffer from being very self-centered. We talk about intelligence, but it basically refers to *our* intelligence. There are other species on this Earth that are intelligent in their own way. Different intelligences. I do not know what will emerge, but it could well be a different kind of intelligence that will be global and have all these properties, making it just hard to predict. We do not know how that will emerge or what the ethics of that artificial intelligence will be like.

W. Brian Arthur: That suggests a line of research I would like to see more of.

There is a classic paper entitled, "What Is it Like to Be a Bat?", by Thomas Nagel.[2] As you have said, we tend to think we are the only ones who can think properly. If cats had the level of intelligence we have, they might look down on us and say those guys do not know how to scratch properly, they cannot catch rats, and so on. I do think that this would be a good counter to the arrogance of human beings if we widen to see that there are other intelligences.

Eörs Szathmáry: There is one thing that I want to remark here, and that is that teaching is extremely rare in the animal world. Teaching is different from learning. For example, in chimps, there is only one rare documented type of teaching where the mother teaches the kid how to crack a nut with two stones. One is used as an anvil. The other is used as a hammer. That took about two years to accomplish the teaching process. That is just one case, after two years. Now compare this to our docility that we have. There is a crucial difference and that is part of our basis of society.

Terrence J. Sejnowski: Something interesting about that observation is that, I have seen films of that process and here is what you see. You see the mother chimp with the anvil, cracking the nut. And the little baby chimp is really interested, and he is focused. But never does the mother chimp take the hand of the baby chimp to show the baby chimp how to hit the nut. This is because they do not have something we have, which is shared attention. Shared attention is when the mother and the child are able to both focus on the same object, at the same time. That is a very important critical step to being able to educate. It is that process.

W. Brian Arthur: A very large part of the economy is autonomous. A lot of the financial part of the economy has humans make some judgments, but the trades are executed autonomously by algorithms. I got fascinated a few years ago when I learned that Aspen trees are a single organism. In an Aspen forest, the trees are sticking up above the ground, but the roots are all interconnected. My metaphor is that what's sticking above the ground in the economy are the parts that we humans see: the screen in front of the traders, and then some traders make something happen.

In fact, what triggered this idea in my mind was that I remember going into the San Jose Airport. I put my frequent flyer card into a little slot and I notice there was a three or four seconds delay before I got my boarding pass. I thought to myself, why? So I put my boarding pass in, and that is the tree above the ground, but below the ground, a lot was happening. Somehow my name and data was captured. That goes to a central computer, through the TSA (Transport Security Administration), the security people, to the UK. These are all of these servers talking to servers, comparing, gathering data. That is already the way the economy works.

There is an underground layer of the economy, which I call the autonomous economy. It is already here, and it is growing. I made some back-of-the-envelope calculations. It is hard to say, but I reckon by the year 2030, it would be similar in size, dollar terms, to what I call the physical economy of the ground. It is already here, and it has autonomous intelligence, unlike human intelligence. Rather it has the ability to take in a situation and react appropriately. According to my base definition of intelligence, there is autonomous intelligence already.

Terrence J. Sejnowski: It could get to a stage where AI could issue the orders to build more apartment blocks in a city, without humans being involved. But we would still be responsible for the goals. We have to specify what the cost function is, which by the way, is essential to AI right now. The cost function tells you if you are doing better or not. Otherwise how do you know if you are learning something that is going to help you in the future?

This raises a bunch of interesting questions. I mean who determines what the cost function is? This came up at a recent meeting when someone was saying, "There is going to be social injustice because right now we have rules about not being biased against certain racial groups when we make decisions for loans, scholarships and so forth." If you are going to be creating artificial intelligence that is going to optimize some goal like wanting the population to do better, it is not going to incorporate these societal, ethical concerns. So how do you put that into the cost function? How do you design it even though you do not know what the unintended consequences are? You have to specify it.

Tor Nørretranders: In terms of the cost function, this autonomous economy, will that mean that it is really an autonomous system of computers that will determine the cost of building the apartment block?

Eörs Szathmáry: I would like to say that this issue about fitness or cost function in designing these autonomous systems is indeed very important because there is an interesting lesson from evolutionary robotics. There have been many well-documented cases where you gave a cost function, and the robot solved it by completely different means. Now, this is important. The system can be extremely opportunistic, and that is a danger.

W. Brian Arthur: I have a general rule in AI. If you can imagine it, then it is here already. Let me give you an example. An awful lot of architecture at the highest level is done in the United States, say in a small elite architecture firm in Seattle. Quite recently, a lot of the detailing and the actual lower-level, brick-by-brick work were done computationally. These are real examples with Computer Aided Design (CAD) and a lot of that is done in Hungary and Eastern Europe because they are technically advanced, and they have people who know how to do software.

If a building needs to be stretched by 25 meters, human architects can do all that, but they will need about three days to prepare new drawings and costing, and to check with building regulations. Now, the people and software in Hungary can do that within three, four seconds, and send it back to Seattle. This is the way aircraft are already being designed, both for Boeing and Airbus. It used to involve human engineers: one person does the avionics, and then passes it to somebody else who says, "You have encroached on my space here." It used to be that the design people had to talk to the construction people and they were starting to toss the designs over the wall. The construction people would say, "Are you kidding, how do we ever make these?" All these have been automated. This is already happening.

Terrence J. Sejnowski: There is a beautiful example of that, which occurred back in the 60s and 70s. This was our early days of integrated circuits. It used to be an art form in the sense that the engineers were trained to generate memory cells and circuits. The idea was to pack as many as you can onto the real estate that you had, and back then, you could do thousands of circuits. As we are getting bigger and bigger, we are staying here longer and longer. So Carver Mead had this brilliant idea. Let's write a computer program that can design the chip.[a] The chip is the computer. The computer is now designing itself and the trade off was that it was not as efficient as the human. The human can really figure out the special ways of jiggering a chip, whereas the computer would just put them in blocks. But at the end it saved so much human labour. It was by far the most efficient way and it took over. All of the design was done automatically by the computer, just like how buildings are being designed in the earlier example.

But the general principle is that once you have this cycle where the computer can take over designing one level, you can then have a hierarchy where maybe a human is making decisions at a higher

---

[a] Ref. 3, p. 1631

level. But why stop there? Let's have that be taken over by another program, which will be on a higher conceptual level.

What you will end up with is a higher hierarchy and it is exactly how our brain has evolved: with the whole hierarchical set of control systems.

References

1. Einstein, Albert. "Einstein at 70." Interview by Alfred Werner. *Liberal Judaism* 16 (April–May 1949): 12, Einstein Archives, 30–1104.
2. Nagel, Thomas. "What Is It Like to Be a Bat?" *The Philosophical Review* 83, no. 4 (October 1974): 435–450.
3. Mead, Carver. "Neuromorphic Electronic Systems." *Proceedings of the IEEE* 78, no. 10 (1990): 1629–1636.

11. Technology and Stability

Eörs Szathmáry, Martin J. Rees, Terrence J. Sejnowski, Tor Nørretranders & W. Brian Arthur

Martin J. Rees: Well, I think these technological advances are exciting prospects and they are changing faster than we can imagine. What worries me is our vulnerability.

If an automated surgery goes wrong, it is bad luck for you, but not for everybody else. On the other hand, for technologies that operate on interconnected networks, there is a vulnerability which could cascade almost globally, and so I do worry about whether we can ever have a system where we can avoid breakdowns.

We already know how electric power outages are more severe now than they were in the past because the city is completely paralysed without electricity.

I wonder if we can ensure that there is not a kind of risk that will be catastrophic and set humanity back. My main concern is that the requirements for reliability are going to get higher and higher as technological systems get more and more interconnected.

W. Brian Arthur: I think there are two things that go beyond the concept of instability.

There are a couple of things that really count with human beings. One is the possibility that if you start to examine societies that depend on economies worldwide, those economies in turn consist largely of technologies, primitive or otherwise, such as fishing boats, manufacturing factories, the Internet, etc.

What are the chances that somehow all our technologies start to collapse? We go back to where we were, say, in the Middle Ages or before that? That is what I would call a true collapse, or that is one sort of collapse. Human beings might still be here, but it is like one of these apocalyptic scenarios. Somehow the Internet goes down, the economy goes down, and there is an awful lot of horrible destruction, including destruction of other human beings or other species. Then a small group of people, whoever is scattered around are left pretty much in the Stone Age or before. They have to reconstruct. I think that is one source of collapse where somehow there is enough disruption that it destroys technologies.

I used to be a demographer. If you look at mortality, it does not vary that much even in wars. What really destroys a civilisation is when its underlying means of support collapses, e.g. the growing potatoes in Ireland. When that collapses, civilisation pretty much goes down. If it is just a matter of waging traditional wars, you see some blips in mortality. But it recovers quickly because of technology. They all come back very fast.

The other thing to think about is the real possibility of very severe pandemics. We are hyper-interconnected now. If something breaks out in some part of the world, especially a pandemic that we have not witnessed, or seen, or properly understood, that can spread very fast.

I am afraid that there will always be disruptions, petty dictators, wars, and we seem to recover reasonably fast. Take the American Civil War. It was quite brutal and really destructive of young men. But ten years later, America is back.

I was always struck by how Germany was defeated twice in a few decades and she bounced back each time. I was working in Germany as early as 1962, and it was more advanced than Britain.

So my question then was: what would it take to destroy the technological infrastructure on the planet?

The answer is: quite a bit. We have to be careful not to be monocultural in terms of everybody depending on a few same technologies.

What would it take for a really major pandemic? I recently studied the 1918 Spanish Flu Pandemic, and nobody quite knows the statistics. Nobody went out and measured it, but there is reasonable evidence that up to 10% of the world's population were stricken and died. Now, that is huge. We can conceivably have something worse than that.

The Bubonic plague was worse. It varied from country to country, and from little town to little town. If you look at Europe in some places it was up to 50%. In other places, it was almost bypassed and might have been 10% or not at all. The reason was that it was carried through various factors including human beings.

But what is very interesting about plagues, like the Bubonic plague, is that the economy is altered for hundreds of years afterwards. There have been very good historical studies of what happened in England.

I think a third of the population in England disappeared in 1349 for two or three years after that. But by the 1500s, things had fully recovered, but there was a scarcity of peasants and workers for a good 100 years after that. This altered the balance of power between the very low-level peasant farming population and the people who had more sustenance. So pandemics could have very severe consequences. We have had a few technological collapses. Beyond that we seem to recover. But I am not trying to be too cheerful about this.

Martin J. Rees: But I think it would be different now for a number of reasons. First, I mean obviously the fatality rate tends to be worse in megacities and places like that. Perhaps the fact that everyone has mobile phones would amplify the disaster by panic and rumour, or, in principle, it could be used to offer help.

But I think there would be real social disruption in advanced countries if the fatality rate got to even 1%. Once the casualty rate reaches levels where hospitals cannot cope, and people realise they are not getting the kind of treatment because of the vast numbers, there will then be a real social breakdown.

W. Brian Arthur: What has changed at the moment is that there is a natural route for migrating birds from the Arctic, North-East Siberia down to China, particular Guangzhou province. Birds migrate. When they go up to the Arctic, they wade around in these shallow pools. Everything that is biological — all sorts of bacteria and viruses — are shared, and these birds come back down. This has been going on for a long time. What is new is that the factory farms in Guangzhou province are right next door to these wild bird populations. The level of technology there is fairly high, yet the level of hygiene of those farms is not high. And since these wild birds are very similar genetically to domestic birds, diseases could be passed. So what we are seeing is not just jet airliners where you have something in Hong Kong one day, and it is in Toronto the next day.

But if you look at the bad pandemics, they are usually from animal species to other animal species to humans. Certainly this is true for Ebola. If I have it right, it comes from gorilla populations and this spreads in Africa. This has been true for AIDS and for many of these things. What is new is that we human beings are clustered together in cities, and we are clustered close to animals, very often to herds or flocks of animals. So the whole system is far more connected.

We have a much more interconnected network. There is a theme in network science and complexity called, "percolation phenomena". If a network is not highly connected, then you can get one domino and these are connections. One domino hits another and hits another and then there is nothing to hit.

Terrence J. Sejnowski: It is actually a phase transition, which in physics happens when you have liquid to ice or steam. What is interesting is that as you vary the density gradually, there is a sudden transition where suddenly it becomes just like wildfire. It just affects the whole population, so you have to be very careful.

W. Brian Arthur: I believe that in many ways we have reached that density because we are far closer to domestic animals than we were ever before, in particular, the last couple of hundred years. Population densities are rising and the rate of human transmission is rising. From a perspective of pandemic risk, I believe we are less safe than we think. We tend to think we are about as safe as we have been in the past 50 years.

Tor Nørretranders: But in terms of infrastructure are we safer than we were?

W. Brian Arthur: I do not know. Let me just say a word on that. There is a problem.

The infrastructure, how we build and use our economy, the means that we are using for support are more and more sophisticated. That means we are more dependent on other sets of technologies. For example we can build fishing boats, or we can have very fancy navigation systems where we can find shoals of fish quickly and easily. If some of those technologies start to go down, we would be back to a birch bark canoe type of era. Those are robust technologies. They are primitive technologies but they are much harder to wipe out. If you have a horse and cart, your horse and cart is likely to do well no matter what. If you have a fancy Ferrari, which bangs up against the wall, you cannot fix the Ferrari but you can do that with a horse and cart.

Tor Nørretranders: If you have intelligence it is likely to work for long time. But if you have artificial intelligence, will it be vulnerable?

W. Brian Arthur: Vulnerable in that sense that if there is major trouble, the first thing to go is highly sophisticated, complicated arrangements.

Terrence J. Sejnowski: There is an interesting trade-off, which is exactly as you described it, which is between efficiency and robustness. What has happened is, as we have gone up the slope toward more efficiency, we are losing robustness. You can go from the Ferrari to the jet in terms of getting faster. You can go beyond from simple civilian jets to military jets. It turns out that human beings can no longer fly military jets in the same way that human pilots control the ailerons and so forth. It has to be done by machine, because of the reaction time. There is an unstable region, and the human is not fast enough to control it. So all these automatic control systems are flying it. The human gets a few nudges here and there, to say I want to go right or left. But without the electronics the thing would fall from the sky because they are not designed to even glide. So the question is how far up that slope have we created these interconnections within our civilisations?

Tor Nørretranders: The Sirius in Northern Greenland were people on sledges, who patrol an area that is inaccessible to snow mobiles and other stuff because there is no gasoline station out there. There is no repair or anything. They rely entirely on old Inuit traditions of using ropes to put the sledge together and so on. These are proven technologies that have been there for a very long time that can be handled by an individual. If you take this microphone probably no individual person on this planet could put that one thing together, because of the division of labour and many people involved. For modern technology, you need other people and you need an infrastructure. With old technologies you can just repair your sledge when it is broken, as they do in Northern Greenland.

W. Brian Arthur: Things that are not that complicated tend to be much more robust.

Things that are very complicated technologically tend to be highly efficient. If high efficiency takes a hit, it all goes down the ladder. But if there's famine due to a bad war, there will be bartering, they will grow their own food.

Eörs Szathmáry: Let me start with a phenomenon that is very interesting. If you look at how ecological systems work and make a model of it, the astonishing result is that, from a mathematical point of view, chaos should be typical. Whereas if you look at real ecological systems, chaos is extremely hard to find. There are only a few well documented cases in ecology. Why is that?

Now, one of the most important things is that it is not one bathtub for everybody. These populations are specially located with limited migration between them. So if you simulate this system you will see there will be many fluctuations. But there is basically zero chance that the whole population would start to do this chaotic behaviour, which is very dangerous for several reasons. Extinctions happen but they are local extinctions. You will be immediately repopulated from a neighbouring local population, which is out of phase with the neighbour. That is a very important lesson.

I raise this issue because this is one of the reasons why we have global stability for many real life biological populations. What you find in the financial domain is extremely dangerous, because with globalisation and the speed of transfer, you basically generate this one bathtub problem for the entire world. Therefore, any perturbation can sink it practically. I know that there have been ideas to prevent them.

But the point is that monocultures are extremely dangerous because they are very prone to epidemics. Another aspect that I am very worried about is the cultural monoculture we are creating via the Internet. If you have a major disaster, I think that potentially there is an asset in cultural diversity because different cultures would react differently to the same global problem. But if everybody has exactly the same patterns of reaction, it means that with time, we might only have one culture. There is a problem because everybody will react in the same way and I am worried about that.

W. Brian Arthur: When the United States was under 13 colonies or maybe 25 states, it was not that well-connected except by wagon or train. You get all these little bathtubs that are not connected. You could run a number of possible experiments. You can have this school system here and another type of system there. Therefore by figuring out what works well in one state, we can transfer it to another state. But all that is starting to disappear.

Tor Nørretranders: But diversity is something that means stability, richness means stability, is that what you are saying?

Terrence J. Sejnowski: Well, the basis of biology is diversity. I mean it is natural selection and you have to have something for it to work on. It is interesting that it makes it robust. It means that regardless of what challenges arise from changes in the environment, there is some small fraction that will survive and continue to evolve. For example, there is a small fraction of humans that are immune to HIV. There is an allele, a receptor for HIV which is missing from these individuals. So they do not respond to it.

Eörs Szathmáry: There is even a subpopulation that has been asymptomatic now for decades, which means that they are in a sense, even resistant to it. You measure the virus and it is definitely in their blood but nothing happens.

W. Brian Arthur: A part of robustness is to have that diversity where there are people or systems that know how to cope with something quite weird and strange. If the planet cooled by 50°C overnight (and there is a movie to that effect), we would still have people and species that know how to cope under such conditions. If you try to make everything the same, it will be highly efficient but not so robust.

Tor Nørretranders: So that deals also with the connectivity issue, because some argue that the Internet allows us to be much more diverse. Many different people can have many different connections. They can find their pen pals across the planet. But other people will argue that interconnectivity means that everything becomes the same and there is no reservoir.

Martin J. Rees: The Internet should broaden the mind but there are books that say how it sort of narrows the mind because, however crazy your views are, you can find someone who shares them. And you do not need to talk to your next-door neighbour who is probably saner than you are.

Eörs Szathmáry: There is one other thing I wanted to mention. I think that what I find extremely dangerous for our stability is when different kinds of instabilities form a cascade. When climate change undermines, for example, the agriculture of a certain region, that triggers many things like civil wars, and mass migration into Europe. When these things start to reinforce each other, because of the positive feedback, even without an asteroid, you could end up in a situation as if you had been hit by one.

Terrence J. Sejnowski: These cascades must have happened many times and as a species, that is part of why we were able to survive these parasites because we can do that over and over again, right? I think that it may not be quite as doom and gloom as you think in the sense that the system has survived many of these catastrophic events.

Eörs Szathmáry: The question is whether or not our civilisation will survive. It is very difficult to wipe out humans as such. But to tax the immune reaction of our civilisation to the point that it gives up, that is another.

Tor Nørretranders: So maybe we are not that much safer after all.

12. Aging and Longevity, Jobs and Migration

Eörs Szathmáry, Martin J. Rees, Terrence J. Sejnowski, Tor Nørretranders & W. Brian Arthur

W. Brian Arthur: There has been a change in the age composition of populations. Two things have shifted radically. Mortality rates have fallen, and it continues to fall. Interestingly, I tend to think it is because we can cure diseases, but the major changes happened in Europe or the U.S. about over 100 years ago. Modern hygiene, certainly in child bearing, has wiped out puerperal fever. Modern sewage systems and clean water are the main things that allow us to live into adulthood where we did not before.

Tor Nørretranders: So, it is more plumbing than medicine?

W. Brian Arthur: Yes. A bit over a hundred years ago, in 1900, you would not want to go into a hospital in Europe or in the U.S., because you could catch an infection and die. You were far better off living on a farm. That started to change in the 1930s, with sulfa drugs and other things. Penicillin came along in the 1940s, and modern medicine arrived fairly late, in the last 60 to 80 years, well, in my lifetime. It has really kicked in, and now we are getting very sophisticated at the genetic level and cellular level as well. But the other thing that has happened, luckily in concert with that, is that fertility rates have gone down. Demographers believe that the world population will likely top at up to 14 billion people total, which we can sort of handle.

Terrence J. Sejnowski: But a lot of those are going to be old people.

W. Brian Arthur: That's right, the age distribution has changed as a result. If the rates go down below replacement rate, which is 2.1 children per family, the age distribution changes. This has already happened in Japan. It is happening in much of Europe.

It is not just the aging population, but there is a kind of boredom, some sort of cultural stasis. I used to work in Austria quite a few years ago. Before the Second World War and during the war, there was little child-bearing for social reasons, nothing to do with the war. It was an older population in Austria. It was a population of people who want dogs for some reason and did not put up with much disruption or noise or music. This was not exactly rock-and-roll. I used to take a train and go to Italy in the 1970s. There were a lot of young people in Italy. The cultural difference was enormous. I don't want to generalise, but there is a difference between the aging population and the younger population in terms of general liveliness, in terms of everything.

It may well be that we are on the threshold of being able to twiddle the genes so that we can live a decade longer. Suppose we could live to be 100, 200, 300 years old. Let me say that with whatever miracle that made it possible, good health goes along with it, there is no deterioration of your faculties, and your body stays pretty well. We tend to ask ourselves, "Would I really want to do that? Would I get bored? Would I really want to play with my great-great-great-great-great grandchildren in 300 years' time? Would I care?"

What we tend not to think about is that other people would be doing this too. A great economist that I admire, Kenneth Boulding, has an essay called, "The Menace of Methuselah".[1] He said, "Well, what if we could live 30, 50 years longer? Hell, what if we all lived to the age 300?" And he said, "This would be very nice, you get nice long lives, we wouldn't be out of here anytime soon." We could watch an awful lot of television and see all the movies we want.

His quip was, "That would all be fine, but how would you like to spend the first 150 years of your career as an assistant professor?"

What we tend to forget is that the rest of the population would be living equally long. Therefore, all institutions would have to adjust. Social institutions, legal institutions, schooling might go on for 100 years, etc. We have not thought about this sort of thing.

Terrence J. Sejnowski: I enjoyed being assistant professor, it was actually a very productive time for me. The problem, I think, is that a 150-year-old professor will have Alzheimer's for sure. By the time you reach 80, the probability of having dementia is up to 30, 40%. Interestingly, Alzheimer's is not correlated with amyloid plaques, which is what the pharmaceutical industry has tried to reduce. There are drugs that reduce amyloid build-up, and the drugs they developed have been very effective in reducing amyloid, but they haven't been effective with reducing Alzheimer's. It is a difficult problem that I think is going to become more difficult as our population ages. It is not just dementia, that's a degenerative disease, but there are many other problems, many disorders that people have. For example, one of the problems with older people is that all their friends die, and they are left alone. Depression becomes a serious problem. There are many serious mental disorders that are going to accumulate as you get older and older. We may be able to cure them or solve certain problems. But we have to start working on it now.

Tor Nørretranders: In an aging population with a lot of very old people, how would the mental faculty of creativity thrive?

Terrence J. Sejnowski: Well, I think the elderly can be creative. Not everybody, but there are some centenarians that are just as sharp as a tack and lead very creative lives, so it is possible. It is not like you are doomed. It is just that things wither away as you get older. In addition to infectious diseases, there are heart attacks and cancer, and these are diseases of the elderly.

W. Brian Arthur: The average lifespan in most places would be 30 or 40. Now it is up into the 80s.

Terrence J. Sejnowski: That's right, that is when all these other diseases start taking hold.

Martin J. Rees: For the people who work on aging, don't they hope that you can actually slow down all these diseases? Or do you think they are doomed to failure?

Terrence J. Sejnowski: One thing we know is that in many cases, there is planned cell death.

In other words, there is a certain program that will kick in under certain circumstances which will force you to die. You have no choice. But that might be subverted.

Tor Nørretranders: What about the biological function of death?

Eörs Szathmáry: A very great evolutionary biologist, second only to Darwin, August Weismann in Germany, was the first person to have thought about it very seriously. The first explanation that he came up with, was that death was good for evolution. It was good for the species, because if you remove the elderly, then of course you are replacing them. It is the young, the new-borns, who are going to introduce the new variation, and the population will be able to evolve faster. He was one of those eminent scientists who returned to the same question again and again, until he realised that he was incorrect. He assumed that the species as a whole is the unit of selection, whereas if you look at the individuals, why should the individual give it up? Why not live longer and produce more children? The first explanation did not confirm the idea of individual competition and selection.

Towards the end of his life, he came up with an explanation which is sort of correct. The older you are, the cumulative chance that you will die anyway will increase due to accidents and diseases. Even if there is no aging, there will be fewer and fewer people, and therefore the force of selection is going to be diminished. The force of natural selection will move toward younger ages, that is one of the mechanisms. That will allow you to, in the long run, accumulate in the population, adverse mutations that become manifest only at later chronological times. That is one of the effects probably at work.

There is another consequence which tells me that you should not raise your expectations too high to think that you are going to have your lifespan extended by 100 years within the next century. I think it is completely out of the question. It goes back to an argument that was made by George Williams, a famous evolutionary biologist. He referred to the example of Henry Ford, the junkyard, and his cars.

The story is as follows. Henry Ford went to the junkyard to look at the old Fords there. He found that the front part of the cars was always good and functional. It was always some other parts of the cars that were not so. He said, "From now on, we are going to produce this particular product in inferior quality," because it does not make sense to keep up the quality of one part if the car dies from other reasons anyway.

Williams argued, "Look, it's actually similar!"

You can assume that because of the process that I described, there will be no reason why evolution would maintain a mechanism that, by itself, would allow you to live 100 years more if all the other mechanisms are declining roughly at the same rate. The expectation is that there will be many mechanisms that go wrong, and on the whole — give or take 10 years — they will give you the same overall decline. Therefore, and this is important, you cannot expect that by making one intervention, all of a sudden you can bump your lifetime up by 50 years. You might, but you will be unable to walk, etc. If you want to bump it up in a meaningful way, you have to do it on all fronts: from the molecular repair processes, mental dementia, the aging of the blood circulation, and all these things.

Otherwise, if you do not improve these things at the same time, they will derail the next step, and the improvements would be done in vain. Because of this evolutionary argument, I think that extending human lifespan is difficult. You can make advances if, by chance, you hit on something which is critical in the network. But again, it is not going to double your lifetime.

Terrence J. Sejnowski: There is a curious demographic phenomenon which is that there is a certain rate of death as you get older. If you make it to 75 or so, the death rate actually goes down. In other words, it is like you have reached the point where you have avoided all the obvious ways of getting killed, and you are one of the lucky ones who has better overall health or organs or whatever. That will take you much further.

Eörs Szathmáry: By the way, that is true for cars. It is exactly the same. Most of the cars will have this death, and then there will be a few which seem to go on much longer than that, because it is a selected sub-population.

Tor Nørretranders: It is also like going from assistant professor to full professor. Suddenly you survive longer. The same is true for technologies. You have technologies that have been around for a very long time, like the knife and fork, they have a good prognosis. But many of the technologies that are very recent, they can have early "sicknesses".

Martin J. Rees: The retiring age of 60 or 65, is probably becoming less appropriate. When people worry about the increase in the aging population, I wonder if they don't realise that one could in fact reclassify people between 60 and 70 as still fully active.

Terrence J. Sejnowski: We have a terrible problem, I am sure that other institutions have it too. We have faculty who have got to old age, and here the question is, how do you review them? At some point, they do become less productive, or no longer as much part of the community as before. But they have reached the point where they are not being reviewed anymore.

Martin J. Rees: I think having a retiring age is actually a good thing. In fact, in my university, we are trying to keep that. I have retired, and I think it is a good system.

Terrence J. Sejnowski: Sydney Brenner once told me, "Terrence J., never retire until you have your next job lined up."

Martin J. Rees: The abolition of retiring ages in professions like academia has had the effect of reducing the number of vacancies for young people. That is one reason why I think it has been a bad transition. The other thing we have to realise, is that although we talk about countries where the population growth is stabilised, there are of course places, especially parts of India and Sub-Saharan Africa, where the age distribution is still very different. The young outnumber the old, and the main issue there, presumably, is to ensure that one does not get a huge population of disaffected young people without employment.

W. Brian Arthur: Where you get chronically unemployed young people, you get trouble. Be it in America among certain minority populations, or the Middle East, or Northern Ireland where I am from. I think the cause of The Troubles (in the 1960s) was in no small part the younger people who were unemployed.

Terrence J. Sejnowski: What did they end up doing?

W. Brian Arthur: Migrating. When I was about 15, I was very well aware that I was on the not-so-advantaged side. You had to pick a country. I thought I would go to Canada. I wounded up in the U.S. instead. Unless you have an outlet for your population, unemployment in general means trouble.

Terrence J. Sejnowski: If migration is an option, it seems to me that this could help the aging population. In fact, in the U.S., although the birth rate has gone down, we do have quite a significant influx of people from Mexico and other countries, our younger population is fairly healthy.

Eörs Szathmáry: I think it is very important to understand what is going on in parts of Europe now. Let's take for example, the case of Hungary. Hungary was invaded by the Turks for 150 years. It has had a terrible effect on the country. Even now on the map, you can see where the Turks have been because the density of the settlements is still much lower than in other regions. It is actually awfully difficult to try to convince the majority of Hungarians that they should welcome those who have exactly the same type of religion and many shared attitudes together with the people who oppressed their country for 150 years. There are very strong historical effects here.

Terrence J. Sejnowski: What is going on that allows a culture to be sustained within populations?

W. Brian Arthur: Where you get tribalism, you get these sharp cultural lines. The real problem in Northern Ireland had very little to do with religion. It was really a tribal thing. Settlers were brought in around the 1600s. Other ancestors of mine were natives and they got dispossessed, etc. You see this again and again, where there are dispossessed populations such as Native Americans or the First Peoples, as they call it in Canada. You see a lingering of cultural and economic effects over hundreds of years, and alongside them is resentment. The romantic side of me says: I love different cultural music, different tribal traditions. But tribalism is slow to go away.

Terrence J. Sejnowski: What about the American experiment? We have immigrants from many different countries. They call it a melting pot, but it is not really a melting pot. Somehow, a lot of these old cultural memories have diluted out with many generations.

W. Brian Arthur: America had the tradition where they said, "We are going to homogenise the people we take in." They could afford to. By contrast, when Britain went to India, they did not say they were going to make everybody English. They said, well, there will be certain people who will be admitted to Englishness, but the rest would be the native population. We will frown a little bit on intermarriage. So, it depends on whether you make an attempt to hold on to differences or make a very conscious attempt to do away with differences.

Eörs Szathmáry: Let me add something more. I think that there is something which you could call a stronger form of tribalism, or maybe not, it goes beyond that. It is a more systemic type of a source of contradiction.

In Europe, they have come up with educational materials for immigrants. Because of problems with language, they tried to put the information into pictures. One of the pictures shows an uneven balance. There is a sign for the legal paragraph on the scale which hangs lower, and on the higher scale, you have different kinds of churches represented. That conveys the message rightly so that in these countries to which you are immigrating, my friend, the rule of law is much more important than the rule of your religion. That is a surprise to many of the immigrants. Can you imagine what kind of tension that generates?

Martin J. Rees: It is certainly changing. Americans will find it harder to absorb people now. The reason is that, in the old days, if you went by ship to America, more so to Australia, that was it. You had very little contact with your family, you were unlikely to go back to see them again. Whereas now you could fly around. You could be on Skype with them every day, you could watch your native country's TV programmes, etc. You do not have the same incentive to assimilate, and so I suspect that more recent immigrants to America won't assimilate in the same way because there won't be the same pressures to socialise since they can keep in touch with their family and friends from their native countries.

Terrence J. Sejnowski: I understand. There are a lot of forces. Assimilation is different in different countries. It is still pretty remarkable that in some places, culture could last for hundreds of years, but in other cases, in a few generations, it's gone.

Reference

1. Boulding, Kenneth E. "The Menace of Methuselah: Possible Consequences of Increased Life Expectancy." *Journal of the Washington Academy of Sciences* 55, no. 7 (October 1965): 171–179.

13. Are We Alone in the Universe?

Eörs Szathmáry, Helga Nowotny, Martin J. Rees,
Terrence J. Sejnowski, Tor Nørretranders &
W. Brian Arthur

Martin J. Rees: I am an astronomer, and astronomy is not only based on physics, but it is also, in a sense, the grandest environmental science. We are observing our cosmic environment which everyone has looked up at through historic times, and we are now realising that it is very complicated. We could trace cosmic history back 13.8 billion years, to a regime when the physics was so extreme, it was unknown.

But at the same time, we are finding out that the cosmic environment is more interesting because each star has planets around it. We now realize that as astronomers, we have to engage with our biological evolutionary colleagues, because we want to know whether evolution happens in these other planets which are rather like the Earth.

The question I am most often asked when people know I am an astronomer is: Are we alone?

Is there life out there? The answer is, of course, we do not know. I get letters from people who think they do know the answer. They have been abducted by aliens or visited them. I say two things to these people: Do they really think that the aliens, having made the effort to come here, would meet a few people, make a corn circle and go away again? I tell them to write to each other and not to me.

But I think it is an important question that I hope we will answer in the next few decades. I think we will understand how life began here on Earth, and that will tell us whether it was a rare fluke or whether it could have happened elsewhere.

This leads to the question: Is simple life widespread? If simple life is widespread, does that mean that complex life is ever widespread? How would we detect it? How would we react if we did detect it?

All we can say as astronomers is that we know that there are, even in our galaxy, billions of planets rather like our Earth, on which in principle life could have evolved. I hope that we will eventually learn how much life there is out there, and if we are very lucky, detect evidence for intelligent life.

Why does it matter?

I suppose it matters a bit because our cosmic modesty would depend on whether we are unique in the galaxy or not. If life turns out not to exist, then we can be less cosmically modest because then our fate affects not just us on Earth, but it may make a difference to the future of the galaxy. But I say that because one thing that perhaps astronomers are more aware of than most other educated people is the immense future lying ahead. Most people are aware of the 4 billion years of biological evolution in the Earth's history, but we are certainly aware that the time lying ahead is at least as long, and that makes us focus on future evolution and post-human evolution.

Most people who are aware of geology and Darwinism, are happy with the idea where the outcome of three or four billion years of evolution, is that the Earth is 4.5 billion years old. But I think even most educated people somehow feel that we humans are the culmination, the end point of all this. They do not realise, as all astronomers do, that the Earth is only halfway through its life. The time lying ahead before the Sun flares up and dies, is as long as the time from start till now. Therefore, the time for future evolution here on Earth and beyond, is as long as the time that its evolution has taken up till now.

Beyond the Earth, the end of the Sun may not be the end of things and the universe may go on expanding forever.

We should not think of humans as being the culmination of evolution. Nonetheless we are humans, so we care about humans. We are certainly an important transition species, the first technological species, and perhaps the species that will trigger the crucial transition from organic to electronic intelligence and life. Our future is important. And that is something that I bring as an astronomer, to concerns of a more practical relevance.

The point I want to make is that the time scale for future evolution, here on Earth and away from the Earth, is as long as the time it has taken for us to evolve. Given that that evolution is going to be auto- evolution (i.e. evolution aided by artificial intervention), that is going to be a lot faster.

I would say that we ought to think about ourselves as being at a fairly early stage in the emergence of the complexity of the universe. But having said all that, I would say that even in this perspective where we look billions of years into the future, as well as into the past, this century is special, because it is the first when one species, namely ours, can determine the future of the entire biosphere.

It may also signal the transition from organic to electronic intelligence, so it is important for those reasons. I would tell young people that they are living through this century, and which scenario emerges will depend on their generation. If they want to be scientists, they should pick the science where new things are happening, new advances are happening fast, or new techniques. Above all, they should familiarise themselves with computational techniques because they are taking over. These techniques are crucial in non-experimental sciences like mine. And they are taking over even some of the experimental sciences.

Terrence J. Sejnowski: You mentioned physics. I think one of the most exciting things that has happened recently, is gravitational waves. Physics is well and alive, and it is on a grand scale now.

Martin J. Rees: Yes, of course. It is an extreme example of how more than 90% of exciting advances came from the most amazing engineering achievements.

Helga Nowotny: It took 40 years.

Martin J. Rees: They have got similar apparatus 2000 miles apart, in Washington state and Louisiana, looking for coincidences. They found a sort of chirp which was indicative of two black holes merging together. They have been looking for this, and it is not a huge surprise. It is a slight surprise that these were two 30 solar mass black holes. The main point is, it is a great technological achievement, and also, probably a firmer vindication of Einstein's theory.

Terrence J. Sejnowski: One of the things about this community, on the paper that came out, a thousand authors were listed.[a] They put everyone on it who was working on it anywhere, because it was a community, so I think that was a good idea.

Martin J. Rees: Yes, so much of science is like that now. It has to be recognised. Giving the Nobel prize to a few people when it is really a group, is not only unfair but it gives the wrong impression of how the sciences are actually done. It is very misleading to highlight individuals. I think prizes should be given only to recognise groups. There are two big prizes which are very good, because they explicitly recognized group-led breakthroughs.

Tor Nørretranders: Despite the millions of planets in the galaxy, it seems that none of them, till this moment, has sent us any signal. Or have we not discovered anything?

Martin J. Rees: We have not. But there are programs to look and I think if we detect anything artificial, like a sort of narrow band signal, I suspect it is unlikely it will be a decodable message. It may just be the malfunctioning or burping of some very advanced machine. But I think the key thing would be to detect something that is manifestly not natural, because that would tell us that technology had existed somewhere else.

---

[a] Ref. 1, pp. 11–13

Tor Nørretranders: But some people say that since we have not found any other civilisation similar to ours despite the great number of planets, it indicates that perhaps the lifetime of civilisations is very short. How do you see that?

Martin J. Rees: I think we do not know anything about how the other civilisations would have evolved. This is a famous argument attributed to the physicist, Enrico Fermi.[b] That argument is, why are the aliens not here? The point there is that there are many stars that are one or two billion years older than the Sun, which could affect planets around them, and that life would have had a headstart on those. Therefore, you might think that the aliens would be here if their evolution is something which happens on many, many planets. So that is an argument against very widespread intelligent life.

But a colleague of mine in England wrote a book with 75 arguments against the Fermi paradox, and none of them are very strong arguments. But I think it is a debatable question, about how surprised we should be that there are no aliens who have manifestly been here. But I think nonetheless, we should look for any kind of evidence for some artificial transmission from space.

We should also look for evidence in any other way. We should look for some sort of artifact in the solar system like in the movie, *2001: A Space Odyssey*, which indicates some aliens have been here or sent some probes here.

---

[b] Ref. 2, pp. 38–39

So I think all these bets are entirely open. I defer to biologists about the origin of life and how likely that is so that will give us some clues. But that still will not answer the question of how likely it is that a simple life can evolve into something we would call intelligent, what the average speed for that evolution is, and also whether that intelligence would need to wish to transmit signals or to travel.

I would personally bet that if we do detect anything, it will be something which is electronic rather than organic and that is an obvious deduction from what has happened on Earth. I mean if what happens elsewhere tracks what has happened on the Earth, then we have had four billion years of biological evolution or thereabout. Then a few millennia of our kind of civilisation leading probably within a few more centuries to machines taking over as it were, and then they will have billions of years. So the civilisation on Earth dominated by beings like us is just a thin sliver of a few millennia in a context of billions of years. If there is any life or any planet out there where life is trapped, what has happened here then will be unlikely to be synchonised with it to the extent that we would catch it in this very short sliver of time, when it is dominated by organic intelligence. Far more likely that it would be far ahead of us, and therefore what we would detect would be something which would be very different from anything organic, anything human.

Tor Nørretranders: It is one thing to detect. But how likely is it that there is a lot of life out there, and would that life be advanced, or civilised, or technological?

Martin J. Rees: We just do not know.

Eörs Szathmáry: Maybe it is a sign of aging. But as I am getting older, I am becoming more and more pessimistic about things like origin of life and such. That is partly due to this experience of how I have been watching the development of research in this. First of all, please do not forget that half of the time of biological evolution of the Earth, which is roughly 3.5 billion years, was strictly prokaryotic bacteria. Our time on Earth is very small.

My prediction (and that is testable in the long run) is that if we found life elsewhere, it would be somewhat similar to what we know about bacteria. Now, whether life as such has come out or not, my hunch today is that because of the many problems that you face in chemical organisation, you have to pass a certain complexity threshold, but also in a certain direction that comes about by chance. You will find this complex chemistry in many planets, but that does not mean that they will be producing anything life-like. Because for that to happen, you need to surpass a threshold, after which it becomes robust and self-enhancing. I could be wrong, but I am not very optimistic.

Martin J. Rees: I agree, and the jump to multicellular organisms took more than two billion years. It is not an easy one, so there may be some that have not made that transition.

Eörs Szathmáry: I fear at the moment that is the case. However, I think it is intellectually very important to ponder about these questions because we are very much plagued by what people call "Earth chauvinism". I look at the solar system now and look at what we used to think about the solar system 20 years ago. Our view has been broadened, and it would be very important — despite what I say — to find examples of independent abiotic life. Europa might be one because of biological descent, but we only have one experiment here. So therefore it is extremely difficult to say what is very contingent, what is a little bit contingent, and what could be completely different.

Martin J. Rees: Absolutely. At the moment, we say that it is so difficult for life to start that this only happened once in the universe. But if we found a form of life say, on Europa or Enceladus, which have clearly evolved independently, that will completely transform the argument. If life can arise independently twice in one solar system, then we could straightforwardly conclude that it has arisen in a billion places in our galaxy.

Terrence J. Sejnowski: What about panspermia? Francis Crick wrote this book.[4] It was an intellectual kind of a foray into science fiction. But it is a serious issue that should be considered, which is that one way in which life might have gotten here is from outer space. Virus-like entities could be embedded in meteorites that showered down on our solar system. And if that is true then maybe, it was found on some of the other other planets. It might be from the same source.

Martin J. Rees: Fred Hoyle used to also believe in panspermia, but he had a better reason for believing in it, because he thought we are in a steady state universe.[5] Therefore, you could relegate the origin of life back to the infinite past. Whereas we know that the universe is only three times as old as the Earth, while the oldest stars with heavier elements in it only twice as old as the Earth.

Eörs Szathmáry: There is one thing that really points against this because presumably if it was panspermic, then you would assume that there must have been a long period of evolution before that. Now the so-called biological or molecular clock is not very accurate. But if you compare, for example, such basic things as the aminoacyl tRNA synthetase, those two words really know the genetic code. They obviously descended from each other. Now if you have an estimate for a divergence of them, certainly even with our inaccurate clock, you could say whether the divergence of this synthetase goes back to something like 3 billion years ago. It is definitely not 6 billion years. What I am getting at is that it gives you a rough idea that indeed it originated within the lifetime of the Earth. That does not rule out that it came from elsewhere, but it rules out the idea that it was due to evolution that was twice as long.

W. Brian Arthur: Looking forward, all we can do is to indirectly see traces of planets on other solar systems. We do not see the planets directly, not yet. But possibly in the next 20, 30 years, we will be able to see directly enough that we can get the spectra of those planets and possibly see traces of some sort of life? Is this correct?

Martin J. Rees: That is correct. I think we could learn quite a bit with more powerful telescopes. Suppose aliens were to look at the Earth with a big telescope, then even if it looked like, in Carl Sagan's phrase, "a pale blue dot", they could learn quite a bit about it.[c] The shade of blue would be different depending on whether the Pacific Ocean was facing them or

---

[c] Ref. 6, p. 8

the land mass of Asia. So they could learn something about the seasons, climate, atmosphere, and maybe even something about the vegetation. That is not an impossible goal if we have big telescopes 20 years from now, and if there are some Earth-like planets within 10 light-years or so, which is again, quite possible.

Tor Nørretranders: Given that there are planets, you can see some kind of change in the atmosphere or something that indicates some kind of life, what then would be our guess at this life evolving into more complicated life? Could we even make a qualified guess of whether that would form cooperation or things like that?

Martin J. Rees: Of course it could be. This is just very simple life that is rather like the Earth while it was two billions years ago. We could see whether or not there's an occlusion in the atmosphere, and that would tell us something. But I think the chance of finding evidence on such planets, that there is a complex biosphere like what we have on Earth would be low, unless it is so advanced it does send out evidence of something artificial. It is unlikely to be a decodable message, but it could be some manifestation of technology, energy generation, lasers, radiowaves, or something like that.

Eörs Szathmáry: Some people have asked, if the universe was teeming with intelligent life, why don't we observe more miracles in the sky?

Martin J. Rees: There is one star, which had variations in its brightness which were not consistent with a mere planet transiting. People do not know how to explain it very much and it was a candidate. I would not bet very much on it, but that was a very interesting candidate.

You are quite right in saying that if we can imagine that there are some planets that have led to a more advanced, a more energy-consuming civilisation or entity, then we might see evidence for it. The fact that we do not see this evidence staring at us in the face is indeed surprising if intellegent life were widespread.

References

1. Abbot, B. P. *et al.*, and LIGO Scientific Collaboration and Virgo Collaboration. "Observation of Gravitational Waves from a Binary Black Hole Merger." *Physical Review Letters* 116, no. 6 (2016): 061102.
2. Crawford, Ian. "Where Are They?" *Scientific American* 283, no. 1 (July 2000): 38–43.
3. Webb, Stephen. *If the Universe is Teeming with Aliens. WHERE IS EVERYBODY?: Seventy-Five Solutions to the Fermi Paradox and the Problem of Extraterrestrial Life*. 2nd Edition. Cham, Switzerland: Springer International Publishing, 2015.
4. Crick, Francis. *Life itself: Its Origin and Nature*. New York: Simon and Schuster, 1981.
5. Hoyle, Fred. *The Intelligent Universe*. London: Michael Joseph, 1983.
6. Sagan, Carl. *Pale Blue Dot: A Vision of the Human Future in Space*. New York: Random House, 1994.

14. Loss of Biodiversity, Ecology, and Economic Growth

Eörs Szathmáry, Helga Nowotny, Martin J. Rees,
Terrence J. Sejnowski, Tor Nørretranders &
W. Brian Arthur

Eörs Szathmáry: After the extinction of the dinosaurs 65 million years ago, we are now undergoing the largest mass extinction on Earth. This is undoubtedly due to us, and the effect that we have on the biosphere, whether directly or indirectly. The data is absolutely shocking. The background extinction rate, which means how many species go extinct through the internal workings of the biota, is estimated to be a mere one to five species per year. The estimates say that the current extinction rate is at least a thousand times that high, maybe ten thousand times that high. This means that literally a few dozen species go extinct every day while we are sitting here. It also means that by 2050, it may well be that up to 50 per cent of the currently existing species will have gone extinct. This really has to be taken seriously. We are in the times of an enormous crisis.

Does that matter? Well yes. The simplest approach is the following. We say that nature and what nature produces has an aesthetic appeal. You can view the creatures in biology as you are viewing pieces of art or beautiful castles and so on. Now, at the minimum, what the data suggests is 40 years from now you might be living in the demolished ruins of a once beautiful castle. That by itself is a disgusting sight and an unpleasant feeling.

But I think the answer is much more profound. The answer is that we simply do not know how much resilience will remain from the original system even after 10–30% of species have gone extinct.

Let me tell you a story: There was a Russian airplane flying between Kazakhstan, and Pakistan. The passengers noticed that the screws in the wings of the airplane were vibrating. Occasionally one of the screws flew away. People started betting whether there will be sufficient screws left before they can land. With the biosphere, there is nowhere to land. You have to remain in flight all the time.

Did the passengers replace the screws every now and then? More importantly, can you replace species when they are going extinct at such a rate? We do not actually know many of them close enough, and we have not genetically sequenced them. So, I think there is a very serious crisis that we have to address.

As to the scientific grasp of diversity, it was not a physicist who took the lead. The first theory of complexity is undoubtedly due to Darwin. The second thing is that nonlinear systems, of which we are so fascinated by, became really important in the golden age of theoretical ecology in the 1930s. It was due to the work of people like Lotka, Volterra, and others. That was triggered by biology and then it went over to different areas of physics. There is a rational way to grasp diversity and to estimate the rates of diversity change. I think that the far more complicated issues are, what are we going to do in order to prevent this incredible prophecy where 50% of the species will go extinct 40 years from now?

Terrence J. Sejnowski: We have known about this for a while. This is not a new revelation. The numbers are pretty shocking. But it seems to me that at least, we have made an attempt in the U.S. to try to do environmental assessments before putting in new buildings and roads. In some cases, they actually look at species that are endangered. But where are these species going? Is it simply where people are? Or could it be due to changes occurring in other parts of the world?

Eörs Szathmáry: There are direct effects, undoubtedly, when people just simply hunt them down. One of the first famous cases was the Dodo in Mauritius. The Dodo was a huge and actually incredibly sweet and funny-looking bird that was killed by people. A hundred years after the invasion of Mauritius, the Dodo and several other species, went extinct due to direct human intervention. Hunting, over-fishing, etc. — these produced a lot of extinction.

Now then, you have indirect effects too. Suppose that there is a nice forest and you kill one of the bird species. 10 years later, the forest starts dying. The seed of this plant has to go through the guts of that particular bird, and only after having gone through, does it make the seed ready to germinate. With that bird extinct, there will not be any new trees. You have cascading effects that are indirect, but in a way that directly follows a logical consequence of human action.

Now comes something from physics. If you think that the biosphere or parts of the biosphere are in what is called a self-organized critical state, then the extinction of any species might trigger a big avalanche. So that is again, a danger because you might think at most one or two species go extinct and that is it. But what if the extinction of one species causes the extinction of a thousand species in a very short time?

Helga Nowotny: I would like to complement what you said by bringing in the one species that has become the dominant species, namely us. If you ever went on a safari to one of the game parks in Africa, it is quite an 'ah-ha' impression you get there. When you see these herds of gazelles that are left in Africa, you get the feeling of what it must have been like to live on the Earth where humans were in the minority. Now we are in the majority. We are growing even bigger, and soon, we will be 12 billion in number. This will lead to the extinction of many species because we are pushing them out.

Martin J. Rees: As to what to do about it, there is an idea from Myers. They are the so-called diversity hotspots.[1] If we try to identify them and protect those particular small areas, that would have a disproportionate effect on cutting down the number of extinctions. Is that still believed, and could it be helpful?

Eörs Szathmáry: Partly. It is related a little to this Noah's ark approach. Now there are two things one has to consider. First, the concept of a minimum viable population size. It is not only one sexual pair. Usually it has to be much bigger, and it depends on the particular population. Only then you can ask: what is the minimum viable ecosystem size? The other thing is that it is a mistake to focus on things like the rainforest. The maintenance of the tundra and the tiger is equally important even though the species number is much lower. So it is not a biodiversity hotspot in the traditional sense, but you could argue that the loss would be even more critical. I think what you should do is not so much biodiversity hot-spotting. But you have to try to preserve a fair number of sufficiently different ecosystems. Whether they are species-rich or not, that is another matter.

Martin J. Rees: I like a quote from the great ecologist E. O. Wilson who says that if human actions lead to massive extinctions, it is the sin that future generations would least forgive us for, because it is something irreversible.[a]

Eörs Szathmáry: This is again a very serious issue. The ecological footprint is basically the need for the wilderness or natural capital that you need per year in order to sustain your existing lifestyle. This includes land, sea, and also that part of the wilderness that consumes the waste that you are producing. As of 2007, we will need more than 1.5 times the size of the Earth to keep it up. The total ecological footprint of humanity is now bigger than the Earth. I think it is dramatic and it is absolutely clearly unsustainable.

Tor Nørretranders: We have problems with diversity. We have problems with the footprint. There are many different suggestions about what to do, like vegetarianism, changing agriculture, or renewable energy. Some people also think that we should somehow interact with the planet and try to do something about the climate problem by sequestering carbon dioxide, a

---

[a]Ref. 2, p. 159

sort of planetary engineering kind of thing. Some people argue that we should use biotechnology to do something about this. Would that be wise?

Martin J. Rees: We know that climate change is occurring. We project that it could be very serious by the end of the century. Not enough action has been taken in the short term. I think the best is to speed up the transition to more economical use of fuels and renewable energy. I personally am pessimistic about anything actually emerging from the Paris meeting despite all the goodwill. I think the best prospects will be to accelerate research and development into clean energy so as to develop it faster and bring down the cost more quickly, so that countries like India are less under pressure to build more coal-fired power stations to work with their growing energy needs. I think the need for more R&D into clean energy should be prioritized.

Furthermore, I personally think that fourth generation R&D into nuclear energy should be done because it has stagnated for the past 30 years. The same design still being used today is 30 years old. I was involved in a campaign to try and increase public funding of R&D, and for political reasons, we left out nuclear because there was no chance that Japan and Germany would come on board if we did that. But I personally think it is good that there is some supplementary funding coming mainly from philanthropists in the U.S. to support fourth generation nuclear power. I think it should be at least explored.

Terrence J. Sejnowski: India just announced that they ordered six nuclear power plants from Westinghouse. So at least they seem to be moving in that direction.

Martin J. Rees: Yes. But those are older designs and we can probably do better. My personal view about climate change is that obviously it is long-term and I doubt that anything very much will be done to cut the annual CO_2 emissions in the next 20 years. But 20 years from now, we will know for sure just what the total climate sensitivity is in terms of how much is the temperature rising for each additional amount of CO_2. We know that, and we have better climate modeling. If the temperature rises rapidly, we are entering into dangerous territory.

Helga Nowotny: There are interesting new developments in Iceland to solidify carbon more quickly. This has been known for some time but now they actually did it. You need basalt rocks, but other kinds of rocks are suitable. You can speed up the process considerably and turn the captured carbon and store them into stones within two years.

It is a major step forward and it is feasible. It is not just in the pipeline: it is actually being done. Once it is done in one place, it can also be done in other places. But you need to get the industry on board in the case of renewable energies, and in particular nuclear. If you have public resistance (and you have it in large parts of Europe), you are not going to get anywhere. It also overshadows other issues where people are much more willing and open to go along. So, it is a very fine line to tread in promoting it.

W. Brian Arthur: One thing we are realising in economics is that if you take a game-theoretic point of view, what we have been trying to do is to get an overall umbrella agreement world-wide fostered by the UN. So somehow everybody is supposed to sign up to some overall agreement, which will tell my country what it should be doing. Of course, there is a lot of negotiation as to how individual countries should be treated.

It turns out that just very trivial economic thinking shows you that it leads to free-rider problems. If I am China, I might say: I did not cause this, and I am going to opt out. Or somebody else might say: I do not believe in this, so I am not on board. It is like if you have a neighborhood that has been thrashed and somebody comes in and tells everybody what to do. It takes months or years to clean it up.

What we are beginning to think in economics is that side payments or side agreement are going to be much more efficient. It is much more efficient for two or three people to get together and say: I will clean up, let us make a little agreement. We will clean up our block and maybe have side agreements. You guys clean up yours, etc.

If you look at what has been successful, such as the agreements between Canada and the United States on acid rain, and more, many of these have worked through local agreements. One country might have been polluting another country because they are downwind, so they make an agreement. Those sorts of agreements historically tend to work.

Overall agreements with some overall body telling people what to do have not been very successful yet. It is getting a little better, but in the U.S., it has led to this massive push back: "The UN is not going to tell us what to do."

There are many possibilities for thinking about carbon emissions and other pollutants. I think we are turning to much more local sets of agreements. None of those are perfect, I admit, but they are a lot better than having nothing.

Eörs Szathmáry: I have a very serious question based on what we discussed. We have this decline in the natural capital and we have economic growth that continues to some extent. Do you really think, or do economists really think that you can sustain economic growth in a finite biosphere forever? Can you elaborate on this issue? I think there is a problem.

W. Brian Arthur: I completely agree there is a problem. Most good economists, and there are plenty of thoughtful ones that I know, do not think that we need economic growth at the expense of the biosphere. I think that might have been true 50 years ago or even in the 1960s. But most of us are by now considerably more enlightened, thanks to people like you. I would not blame the economics profession for putting that kind of thinking out there. It is actually politicians, as I see it. At least I hear this again and again, politically, especially in the United States.

Politicians seem to think that there is a trade-off between economic growth and taking proper care of the environment. I think very good research, be it political, sociological, technological or economic, shows that you can have reasonable amounts of both. But we need to give a lot of thought as to what we are doing. Do we need Soviet-style old-fashioned factories? China is trying to get rid of those. Do we need to just pump up pollution all over the place without cleaning it up? I think we can do both.

A good thing, in my opinion, is that the economy is becoming more and more digital. We have debated whether that is a good thing but one good aspect of things becoming more digitized is that it is using fewer natural resources by and large. If I do not need to send a package anymore I can send an email, then maybe, pretty soon we would not need freighters spilling oil onto the sea, and we will be able to have other forms of energy. Maybe we will be able to have a much lower energy footprint because so many things are being delivered digitally.

I am not saying this is a wonder, but I think we are going in the right direction with the economy. I very much like the idea that is getting through to society and politicians and economists as well, that it is not costless to just pump effluent into a river. We are starting to wake up and that brings me to a remedy.

One thing I want to mention is that I believe what we truly need is a different set of attitudes to nature. Let me put it this way. Many other people and I have studied major difficulties in the past. We had slavery in the United States up until the 1860s. We had wretched conditions and factories in Europe, particularly in England in the 1850s. What seems to happen is a very slow change in attitudes. For example, it is not right to take fellow human beings and enslave them. It is not right to take small children or women and make them work 14 hours days, or men for that matter. So once public attitudes start to shift then you see this other machinery kick in.

I believe attitudes start in kindergarten. I think we can get a great deal forward with just a minor shift in perspective, and in what we teach our children in school. To go back to the famous speech of Chief Seattle, "The Earth does not belong to us, we belong to the Earth."[b] If we can get that idea to five-year-olds, I think that there would be a lot of changes in the future.

Tor Nørretranders: If we take this change of attitude issue, I wonder what does science contribute here? I invite you to reflect on this thought for a moment: Is the attitude of science towards nature, the attitude of science towards human beings (or human societies)? The way science is dealing and talking about nature, is that part of the attitude we need or the attitude we need to have? Is science part of the solution or part of the problem?

References

1. Myers, Norman, *et al.*, "Biodiversity Hotspots for Conservation Priorities." *Nature* 403, no. 6772 (2000): 853–858.
2. Myers, Norman. *Gaia: An Atlas of Planet Management.* Rev. and updated ed. New York: Anchor Books, 1993.
3. Kaiser, Rudolf, "Chief Seattle's Speech(es): American Origins and European Reception," in *Recovering the Word: Essays on Native American Literature*, eds. Brian Swann & Arnold Krupat (Berkeley: University of California Press, 1987), 525–530.

---

[b] Ref. 3, p. 527

15. A Sustainable Population for the Planet?

Eörs Szathmáry, Helga Nowotny, Martin J. Rees,
Terrence J. Sejnowski, Tor Nørretranders &
W. Brian Arthur

Tor Nørretranders: How many human beings can there be on this planet?

Martin J. Rees: It depends very much on lifestyle, doesn't it? To take two extremes, if they all live like present-day Americans, using as much energy and eating as much beef, then probably 2 billion might be the maximum population. If on the other hand, they live in little capsules, eat nothing but rice and live in virtual reality, there could be 20 billion. So, it could be anywhere between that.

However, I think the main concern is the rate of change. In countries where the population is still rising fast, like in parts of India and Sub-Saharan Africa, there is a risk that they will not escape the poverty trap, and that may allow this growth to continue.

It depends on whether they do reduce fertility rates to replacement level. People say that such population reductions are due to women being educated and all that. But I was talking to some Africans, and they said that it would not, because people in Africa like to have large families for cultural reasons. It is not obvious that the population of

Africa will stabilise even if women were educated. This is a sociological question. I do not know but these are big uncertainties.

Helga Nowotny: That reporter did work in South Africa precisely on that question. He found out that attitude changes once men adopt the view that it is better to have fewer children and you can still be considered to be a man in terms of say, tradition and culture. Whether that will happen is linked to culture and economic perspectives, because we know that large families are needed when you have no insurance. It is always a combination of it. It is not just numbers. It is the geographical distribution. It is also the age distribution because once you have the bulk of young people, they reproduce, and you get more young people, babies.

W. Brian Arthur: My colleague, Paul Ehrlich, had this idea that the Earth has a carrying capacity for human beings. Time and again, it has been disproved. We could have 30 billion people, god forbid, and somehow the Earth, or humans would still find a way to get by. It might be miserable, and it might bring irreversibility into the climate, or something worse than what we have now. But it is very hard to define a carrying capacity.

But I do want to point out some things here. Once you get an awful lot of young people in the population, that is something demographers call, "population momentum." Japan, for example, does not have that much momentum. There are not that many young people who will produce the next generation. But Africa has, so we have to account for Africa's population momentum.

There is one other thing that has not been much talked about, and I would like to toss this in. Having children later and spacing them is extraordinarily powerful. I worked at the Population Council in New York for 7 years. One of my colleagues, John Bongaarts, wrote a paper very early on.[a] It turned out that in the 1970s, when China came up with the one-child policy, they had been heavily influenced by the Club of Rome report. Well, not everybody in China, but a few key people were. And so China said, "We need to do something about all this!"

My friend, Bongaarts, showed that they could have exactly the same population in the future, say as they have now, total numbers, if everybody was allowed two children, but could space them, so maybe you get your first child for a woman at age 30, and the second child at age 36, or at least amongst those ages. This may sound draconian, but the one-child policy is extremely draconian. The point is that with better understanding of demography, you can indeed have better achievements, and there are different feasible ways to get them.

Like everybody else in the West, we hope that we have free will, or we are allowed to freely choose the number of children we have. At the same time, that is not the case all over the world. The important factors seem to be if there are good arrangements for people in their old age.

Are women literate? Do men have other ideas than just having women and showing how powerful

---

[a] Ref. 1, pp. 600–601

they are by having a large family? I should confess and say, truth be told, I have four children.

It would be a mistake to think that humans are very good at regulating their own population. This is something called the demographic transition. If your children are going to live to an old age, you do not have to have six, seven, or eight kids.

I don't think families are calculative, and they are certainly not calculating for the good of the nation or the world. But these families are finding alternatives to having a large clan following them. The social arrangements become slightly different. I am sure every case differs, no doubt. But this has been what has gone on for the last 150 years.

Terrence J. Sejnowski: Interestingly, the U.S. had this demographic transition about 100 years ago. China did the whole thing in 10 years. It is astonishing how that happened. There is another important thing to think about, which is the size of cities. In China, it is not unusual to find cities with 10 million people. It astonishes me the degree to which humans can adapt to the most adverse conditions, in terms of both food and quality of life. In a sense we are a robust species. But I am not sure that that is good if you want to be able to limit the population, because it means that whatever happens, 14 billion people will adapt to it. The quality of life is really going to go down.

W. Brian Arthur: Population size is an issue, urbanisation is an issue. But in my opinion the biggest issue to do with population is actually a geographical one. There have been population settlements. For example, 2000 years ago there were more people in Northern China than Southern China because there was a disease gradient, things like malaria. It was very difficult to live in a hot and moist climate. The population tended to settle, or at least be concentrated in much bigger cities in the North.

The population around the world, has now spread out geographically. In Bangladesh, people have moved into a place called, the Sundarbans, which is regularly flooded by rain or by tidal rises. You can think of populations in different regions as being where people are comfortable, and what works and supports that sort of population. It is very ecological or biological.

Then along comes climate change, and that is not going to work evenly. Recently, I asked my sister in Ireland, "What do you think of climate change?" She says, "We love it, it's incredible! You can sit outside and drink coffee now."

The change in environment is not even, and there is going to be major geographical redistribution in population. That does not just mean refugees. It could also mean wars.

I think that the main issue to do with population is the way it is geographically distributed. It has been repeatedly said that there's little that we can do about it: neither science, nor the government. But I think what we can do about it at least is be aware of what the major issues are. Urbanisation is a big issue. Running cities properly is a big issue. But the geographical distribution is going to bring problems, and those problems are not going to be easy.

Reference

1. Bongaarts, John and Greenhalgh, Susan. "An Alternative to the One-Child Policy in China." *Population and Development Review* 11, no. 4 (December 1985): 585–617.

16. Ethics and The Relevance of Science in the Public Sphere

Eörs Szathmáry, Martin J. Rees, Terrence J.
Sejnowski, Tor Nørretranders & W. Brian Arthur

Terrence J. Sejnowski: Here is the problem — no one can predict what impact a discovery is going to have, whether it is going to be used for good or bad. There is no way you can know.

Eörs Szathmáry: Exactly, and this is true for both the positive and negative consequences. This is why the relevance business is actually nonsense, because you cannot forecast the most exciting technological developments.

W. Brian Arthur: How can we tell the importance of research that is getting proposed? You can't tell, nor should you tell. You don't know what the so-called relevance might be down the road. The relevance itself might not have been invented yet, because it might only become relevant after some other technology is in place.

Terrence J. Sejnowski: This raises the issue of the delay between a proof or principle, and it becoming a practical, relevant part of technology. The delay is typically 20 to 40 years.

Martin J. Rees: It is true that when a pure discovery is made, we do not know the balance of good and bad. But when you get to a particular application, then of course it is analogous to the medical Hippocratic Oath. Do you make bioweapons, do you make hydrogen bombs, and all that?

In that context, it is a real question. It is still a question of the extent to which scientists should put their ethical views above the ethical views of the rest of the public.

Terrence J. Sejnowski: Well, there is anguish. I know people who are really concerned and worried about what their research is going to be used for. On the other hand, there is another point of view, which is that some things are so sweet that they are going to happen anyway.

Martin J. Rees: There is a lot of commercial pressure as well. My view is that enforcing any such regulation is as hopeless as drug laws or tax laws. We should try, but I think what scares me most is that we are not going to be able to prevent people from misusing all these technologies, by error, or by design. That is my number one fear.

Terrence J. Sejnowski: I think the attempt to come up with regulations immunizes science from public criticisms that we're going into these things without having thought through the consequences.

W. Brian Arthur: If someone says, "What is the importance of this? What is the relevance of this? What is the use of this?", my economist mind says, "What does that mean? Use? Importance? Importance for what? Making better washing machines? Making the traffic flow a little bit better?"

If you went to Beethoven, and he is just about to write a symphony, and you say, "What's the use of this? What is the importance of this economically?" "Well, I can fill up a few more theatres, maybe."

The point I want to make is fundamental, so I am not disagreeing with what you say. Of course, there should be relevance, and there is plenty of relevance in science, but the deepest point is that science exists for science, in my opinion. I hate to bring all of this back to human beings. What is the use of the biosphere for humans, do we care if humans go extinct? But coming back to humans, we do have a sense of wonder, we have a sense of elegance, and we have a sense of something just being, "Oh, wow!"

Martin J. Rees: You can appreciate science without doing it, and that is possible through the media, etc. In fact, more people can actually do science. It used to be that there were rather few sciences, like say, botany, where amateurs could actually make absolute professionals. That can now be done in other fields.

For instance, there was the Galaxy Zoo Project, where 3 million galaxies were classified by amateurs, which was better than what could be done by machines at that time.[1]

You could download the time series observations for one star, and there was an algorithm which found most of the planets for the regular dips. But there was one very special system where the planet had double stars. It is an amateur who had found it, we discovered an exciting new planet! So that is an example of how the ability to access and download large datasets allows far more people to participate in genuine scientific discoveries.

Terrence J. Sejnowski: Yeah, crowdsourcing!

Martin J. Rees: Another example was that they got amateurs to analyse 18ᵗʰ century ships' logs, which have useful climate data on the temperature of the ocean.[2] Some got interested in the rest of the logs' content and began reading the naval history and the details of what happened to particular ships, etc. This is just an example of where the availability of huge amounts of written data and measurements do allow people to participate and find out their interests in a way that they never could.

Terrence J. Sejnowski: There is another example, which was *Spirit and Opportunity*.[3] Steve Squyres, was the head of that project. He talked about what they went through to get there. It was $800 million program. How do you justify that to the public?

He had a real difficult decision to make at the very beginning about when to give out the data. The rovers were collecting all these data every day. The team wanted to hold on to the data for a year or two, so that they could mine it and publish papers (that is how academics work).

Steve Squyres felt that that was too restrictive, and the fact is that since the public was paying for this, they should get the raw data as it came out. So he made the decision. The scientists weren't very happy. As soon as the data came in, it got out onto the Internet.

What happened, he said, was really astonishing. To go from the raw data to something you can actually look at requires a lot of pre-processing. You have to go through and get rid of the artefacts, you have to figure out how to normalise it. There is just a huge amount of work that has to be done. But what actually happened was that the raw data would be downloaded, and he would go to sleep, and the next morning, somebody would have done all that work for them. The work was sped up by several factors.

Again, you are harnessing the interests, eyeballs, and the talents of the rest of the world who otherwise wouldn't have access to that data. Rather than hoarding everything, I think that the more data that gets out there, with more people looking at it, and more tools and techniques analysing it, we will make progress quickly.

References

1. Galaxy Zoo. "The Story So Far." Accessed February 3, 2018. https://www.galaxyzoo.org/#/story.
2. Spatial.ly. "Mapped: British, Spanish and Dutch Shipping 1750–1800." Accessed February 3, 2018. http://spatialanalysis.co.uk/2012/03/mapped-british-shipping-1750-1800/.
3. National Aeronautics and Space Administration. "Spirit and Opportunity." Last Modified August 4, 2017. https://www.nasa.gov/mission_pages/mer/index.html.

Printed in the United States
By Bookmasters